THE UNIVERSAL SENSE

THE UNIVERSAL SENSE

SENSE

How Hearing Shapes the Mind

Seth S. Horowitz

BLOOMSBURY

NEW YORK • LONDON • NEW DELHI • SYDNEY

Published by Bloomsbury USA, New York

All papers used by Bloomsbury USA are natural, recyclable products made
from wood grown in well-managed forests. The manufacturing processes
conform to the environmental regulations of the country of origin.

LIBRARY OF CONGRESS CATALOGING-IN-PUBLICATION DATA

Horowitz, Seth S.
The universal sense : how hearing shapes the mind / Seth S. Horowitz.
p. cm.
ISBN: 978-1-60819-090-4 (hardback)
1. Hearing—Physiological aspects.   2. Sound—Psychological
aspects.   I. Title.
QP461.H594 2012
612.8'5—dc23
2012009642

First U.S. edition 2012

1   3   5   7   9   10   8   6   4   2

Typeset by Westchester Book Group
Printed in the U.S.A. by Quad/Graphics, Fairfield, Pennsylvania

For China Blue and Lance,
partners in acoustic crime

and

Arnold Horowitz,
for late-night weird ideas

CONTENTS

Foreword and Acknowledgments

A first book is a very hard thing to write, especially when it involves something you feel passionate about. It requires bringing together elements scattered over decades of experience and interactions with a huge number of people, places, and things, living and otherwise. Writing about things heard is particularly difficult because so much of the inner life of sound dwells below conscious thought. At first thought, it seems easy—many of the sounds we pay attention to in daily life are words. You can transcribe conversations or lyrics into written form in a straightforward fashion because words are bound by the conventions of language. But go a bit below the basics of the written and spoken word and you find that the rules of written language only give you a piece of the richness found even in plain speech of a nontonal language such as English. If I write "What?" you, as the reader, understand that I asked a question. But imagine hearing me yell it (WHAT!!!!). Suddenly there is an entirely new context based on how I changed the sound—you are interrupting me, I'm angry or impatient, or you just gave me really bad news and I'm questioning the reality of what you just said. Or if I say it very quietly after a long pause (. . . what?), have you just given me really bad news? Just by changing the sound of a simple word, you gain insight into the emotional, attentional, and behavioral state of both the speaker and the listener. While you can describe the sound of a contented cat as a purr, how can you explain how it induces a feeling of calm in its owner or frustration in a person who has become the object of said purrbox's affections despite a massive allergy to felines? Trying to explain these reactions requires forays into amplitude modulation, interspecies communication, and the emotional function of the brain, both human and feline. And what about the squirm-inducing sound of fingernails ratcheting down a blackboard? Why would we have evolved such a visceral response to a very specific behavior that uses a piece of technology that was only invented in 1815?

The way sounds are made and heard (or not), the effects sounds have on your mind, your emotions, your attention, your memories, your moods, are so vast as to be almost beyond description. There are literally hundreds of excellent (and some not so excellent) books covering individual pieces of this vast puzzle. Yet at the heart of all sound, its perception and effects on living things, lies a mathematical heart that ties together the most basic interactions of matter, energy and the mind.

I decided to write this book because for more than thirty years I have been fascinated by all types of sounds. Sound has held my attention and my passion as I have tried to understand it from a variety of different perspectives—from R&B musician to digital sound programmer, dolphin trainer to auditory neuroscientist, music producer to sonic branding designer—and to integrate it into a single theme: how sound and hearing have shaped the evolution, development, and day-to-day function of the mind.

A couple of words about why this book may be different from other books you've read on sound. Most recent science-related works about sound and its various children such as speech, music, and noise are based on neural imaging studies. Techniques such as functional magnetic resonance imaging (fMRI) create beautiful images of what part of the brain is active when listening to, looking at, or thinking about certain things. For the most part these studies focus on the cortex, the huge convoluted volume of brain that is characteristic of humans and other big-brained mammals such as dolphins and chimps, and hence is what's called "top-down"—you look at the areas that are involved in the final processing steps for the activity you're interested in. My take on things is based on how things start in the outside world and work their way in, from the first physical sensations in the ear (and occasionally other sensors) through the lowest parts of the brainstem. My perspective is bottom-up, focusing on elements that underlie and drive high-level cortical and cognitive functions. Both perspectives are critical in science, but to me the bottom-up approach gives a more intimate understanding of the *umwelt*, the world built from your senses. To me (and I hope to you after reading this book), this gives one brain a better chance of understanding the deep processes of another brain in the context of the ever-changing world in which we are embedded.

As with all projects that try to take a broad view of large things, many fine features will get lost in the mix. This book will not be a textbook on auditory neuroscience and perception, but I will try to explain my wonder at how cells a few tens of microns long, working on the picovolt scale and opening and closing molecular channels thousands of times per second, underlie your recognition of your mother's voice. This book will not give any ultimate answer to the biological basis of appreciation of music, but it will try to explain why science has such a rough time addressing this type of basic human behavior. Likewise, it will not tell you how to become an award-winning sound engineer, but it might explain why you really shouldn't shell out any money for ringtones that are supposed to make you irresistible to the opposite sex or drive teenagers off your lawn. And while it won't be able to explain the choices of filmmakers who insist on inserting explosion sounds in outer-space battle scenes, it will take you on an acoustic tour of places beyond Earth in the hopes that some young readers will actually get the chance to hear the winds of Mars for themselves, and perhaps remember that they read it here first.

It has been more than thirty years since I first thought about writing a book about sound, and if I waited another thirty, I still wouldn't have the whole story—and I would have even more seriously blown my deadline getting this to my very patient editor, Benjamin Adams. But in those thirty years, I have been blessed with a wide cast of teachers, supporters, colleagues, and friends who have shared and shaped my passion for sound and listening. The short list includes the late Gerald Soffen, former NASA director of life sciences, who first sparked my interest in astrobiology (and was the first to ask me what I thought Mars would sound like). Martha Hiatt of the New York Aquarium was the first person to ever successfully teach me patience by holding out the carrot of letting me work with and learn from Lily, Starkey, and Mimi, the dolphins who inspired my love of animal behavior. Peter Moller of Hunter College tricked me into loving psychology research and introduced me to the world of neuroethology, as well as suggesting that I follow it as a career, starting at Brown University. There I met my advisor, Andrea

Megela Simmons, and my postdoctoral advisor, James Simmons, as well as some amazing teachers and students who became colleagues, including Sharon Swartz, Barry Connors, David Berson, Diane Lipscombe, Judith Chapman, Rebecca Brown, Mary Bates, Jeffrey Knowles, and others too numerous for my page limits. Peter Schultz, friend, scientific provocateur, and director of the NASA Rhode Island Space Grant, has encouraged and helped fund some of my strangest projects. The late, great, and much missed Ed Mullen, the engineer extraordinaire who dubbed me "Dr. Evil" for my interest in the "weirdest things ever built in the name of science" (e.g., a bat-mounted laser pack), helped me carry out some of my most fun studies. Rachel Herz, fellow sufferer of nonsense at Brown and author of *The Scent of Desire*, got me into this whole book-writing mess. Layton Tolson, who has provided constant cheerleading and enthusiasm, also helped me in my first forays in clinical studies of sound and sleep. I also am deeply grateful to Brad Lisle of Foxfire Interactive for bringing me into the *Just Listen* project and giving me the opportunity to hear and work with Evelyn Glennie, an amazing percussionist who gave me some wonderful insight into her unique perceptions of music. Mary Roach, one of my favorite science writers, held my hand through a lot of the first-draft temper tantrums. I also gladly thank my agent, Wendy Strothman, and my editor, the aforementioned Benjamin Adams of Bloomsbury USA.

My greatest thanks have to go to four people who are in some ways the most to blame. My mother, Marlè Fishman, insisted that whatever I do, it should take me where I want to go. My father, Arnold Horowitz, who is my intellectual and creative hero, drilled into me from a very early age that if someone says something is impossible, that just means it will take a bit longer to achieve, and that the only real failure in working with equipment is if it fatally electrocutes you. My friend and partner in sonic crime for almost fifteen years, Lance Massey, the first person to get a degree in electronic music from Oberlin, the creator of the T-Mobile audio logo, and the man with whom I have been trying to create weapons-grade sounds for over a decade, has kept me going through all this by reminding me that no matter how much science I throw at the writing, in the end it's all just a jingle. And last but not least, my beautiful, brilliant wife, China Blue, sound artist extraordinaire, who has not only put up with me but collaborated with me on projects that anyone else would have had me locked up for, and only argues with me over access to the soldering iron. Thank you all. This is your fault.

INTRODUCTION

IN RECENT YEARS I spent a lot of time in my office in the basement of the Hunter Laboratory of Psychology at Brown University. I was surrounded by the effluvia of ongoing experiments: little backpacks that allow bats to carry lasers, middle-ear bones from a rhesus macaque, photos of my beaming wife curled up in the Vertical Gun chamber at the Ames Research Center, and calibration instruments so numerous even I don't know what they are all for. Amid all these distractions, I was often glad it was so quiet. Most people yearn for an office with a window, but I rather dreaded that possibility. While I occasionally had to put up with an alarm indicating that the shop had flooded again or a temperature control had died in the frog colony room, or the occasional beep-beep-beep of a psychoacoustic experiment being run in the lab space across the hall, in general I liked the isolation, and the feeling that I had time to think. Time to think is more of a restricted commodity than you might expect in an Ivy League science department—most time seems to end up devoted to writing memos about the fact that the aforementioned temperature control has failed five times in the last two weeks.

Which leads you to wonder: Is it really quiet in these halls? Even without any alarms or graduate students trying to debug experiments, if I actually paid attention, I could hear pretty much constant noise: low-pitched rumbles of the air vents vibrating, traffic driving past, voices in the next corridor over, humming from an elderly fluorescent light fixture over my head, weird gurgles from the old mini-fridge next to my desk. If I pulled out a sound level meter, the room's noise level would probably be about 45–50 decibels (re: 20 µPa sound pressure level [SPL]—an important part of the definition of loudness but one that is almost always left out). Yes, that's on the quiet end of our normal experiences, but it is certainly not silent. It's quieter than a gunshot, certainly, but if I were a bat or a mouse, the hum from the fluorescent bulb would be a screamingly loud 80–90 dB SPL. And if I were a frog, the gentle vibrations of street traffic would feel about as powerful as an earthquake of about 5.0 on the Richter scale.

There is no such thing as silence. We are constantly immersed in and affected by sound and vibration. This is true no matter where you go, from the deepest underwater trenches to the highest, almost airless peaks of the Himalayas. In truly quiet areas you can even hear the sound of air molecules vibrating inside your ear canals or the noise of the fluid in your ears themselves. The world we live in is full of energy acting on matter—it's as basic as life itself. And the reasons the constant thrumming doesn't drive us all insane are the same reasons we get distracted by radio jingles and can't read when the TV is on: we are good at choosing what we hear. But even if we don't hear a sound, someone else does.

The natural world is based on variation, even among vertebrate animals who tend to share a lot of evolutionary history.

There are plenty of blind animals: cave fish live beyond the reach of light in underground caves and map the world by changes in water flow around their bodies, and the muddy river home of Indus river dolphins blocks most of the light, leaving them to rely on echolocation and the odd habit of swimming on their side to feel what is in the mud. There are plenty of animals (including us) who are unable to detect the songs of electric fields that underlie so much of the behavior of electric fish and the hunting behavior of sharks, or the beauty of the ultraviolet world as seen by bees. Plenty of animals have very limited senses of smell (humans again); animals such as the armadillo have a limited sensitivity to touch; and we can but hope that vultures have a limited sense of taste. But here's one thing you never find: deaf animals. Why?

Let's step back a moment. There are a lot of organisms, little one-celled things with no nervous system, that couldn't be appropriately described as having a sense of hearing. And there are lots of animals, even household pets, whose hearing has deteriorated with age or been blown out by some untoward event. So, speaking more specifically now, there are no normally deaf vertebrates. And that distinguishes hearing from all the other senses we know, including sensitivity to electrical fields and ultraviolet light.

So why do all animals with backbones hear? Or to put the question another way, why is hearing the most universal of all senses? And if it is so crucial a sense, why do we humans so often ignore it at a conscious level, unless we're trying desperately to block out the noise of the subway or checking out the latest music downloads?

Sound is everywhere. From the night chorus of frogs in the deepest rain forest to the emptiest wind-blown stretch of the

Antarctic, you are surrounded by, embedded in, and molded by sound and vibration. Anywhere there is energy, including the depths of intergalactic space, is a vibratory region. Some are richer than others, but none are totally silent. The range of measured vibration is immense. At the top of the spectrum is the insanely fast 9,192,631,770 cycles per second of an energized cesium-133 atom. Near if not at the bottom are the gravity-wave-induced pulses of the sound of black holes (a B flat 57 octaves below middle C, according to Andrew Fabian of the Institute of Astronomy in Cambridge). But these vibrational extremes are not of biological relevance to us—everything our brains can handle happens in a more constrained time frame, from hundreds of nanoseconds to years, far from either end of the spectrum. Living things are tuned to pick up information of interest and use to them, signals that will let them gather information about the environment, their friends and family, their predators, and their alarm clocks. But even within the limitations imposed by biology, there is a tremendous range of information to be gathered, whether by human hearing, bat sonar, or the head-knocking communication system employed by naked mole rats.

Vision is a relatively fast-acting sense that works slightly faster than our conscious recognition of what we see. Smell and taste are slowpokes, working over the course of seconds or more. Touch, a mechanosensory sense, can work quickly (as in light touch) or slowly (as in pain), but only over a restricted range. By contrast, animals and humans can detect and respond to changes in sound that occur in less than a millionth of a second and to the content of complex sounds over the course of hours. Any detectable vibration represents information, to be used or ignored. And in that simple concept lies the entire realm of sound and mind.

Whether it's a humpback whale listening to hours-long song cycles during its migration or a bat using a submicrosecond difference in echoes to determine if something is an edible treat or a branch to be avoided, sound helps animals find food, mate, play, and sleep; ignoring it can get them eaten pretty fast. Which is probably why vibration detection, including what we humans with our fair-to-middlin' ears call hearing, is one of the most basic and universal sensory systems that any earthly organism can have. What is detectable, what is discernable, and what is relevant are the bases for parsing out raw vibration into silence, signal, and noise.

All research into hearing seems to home in on two basic facts: (1) if a channel of information is available, it will get used by living things, and (2) sound is everywhere there is life (and other places). Anywhere there is matter and energy, there is vibration, and any vibration can transfer energy and information to a receiver who is listening. And the wide range of vibration perceivable by living things, from the single thud of a footstep that shuts up a frog chorus to the incredibly high-frequency sounds that form a dolphin's natural ultrasound, requires a sensory system thousands of times faster than its slower cousins, vision, smell, and taste. It is this faster-than-thought auditory speed, with a wide range of tones and timbres that visual color cannot hope to match and greater flexibility than the chemical sensitivities of taste and smell, that lets sound underlie and drive a fantastic range of subconscious elements in the living organism. Combined with wildly divergent ways of listening by different species and the increasingly complex ways of using information by living things, the presence of sound drives the evolution, development, and day-to-day function of the mind. How it does this is what the rest of the book is about.

Chapter 1

IN THE BEGINNING WAS THE BOOM

IN THE SUMMER of 2009, my wife and I were invited to go to NASA's Ames Research Center to try out some sound recording work on the Vertical Gun. The Vertical Gun is a 0.30 caliber light gas gun that fires custom-made projectiles (ranging from ice to steel) into targets at incredibly high speeds, up to 15 km/sec (about 33,000 miles per hour, or around fifteen times the muzzle velocity of an M16 rifle). It is used to simulate meteorite and asteroid impacts. The gun's force comes from a powerful explosion, about half a pound of gunpowder detonated in 50 liters of hydrogen gas.

The bright red barrel is three stories high, pointing down at the ground through a sealed central chamber—imagine an elevator shaft that you really, really don't want to stand in at the wrong time. The gun's angle is changed by an old Nike missile launch elevator and fired through one of four ports at different angles into the central chamber. The main chamber, painted a cheery sky blue with walls thick as battleship armor, can be brought down to vacuum levels near that of outer space to simulate impacts on airless bodies such as the moon or filled with custom gas mixes to replicate an atmosphere like the Earth's. The test chamber is about 2.5 meters around with a central "target" that can be filled with sand, water, ice, or any other substance that might do something interesting when hit by a projectile at tens of thousands of Gs, and has a variety of ports to allow images of impacts to be captured by stereo video and thermal cameras with frame rates of up to 1 million frames per second. The walls of the room in which the chamber is located are lined with targets and projectiles from previous experiments— Lucite, Lexan, glass, steel—all showing the cratering and shattering effects of high-speed impacts. The big red button that fires the gun is safely in another room.

We were invited by Peter Schultz, a friend and colleague from the Brown University Planetary Geology department who is one of the world's experts on crater formation. He can wax eloquent on rates of crater formation to help determine the age of planetary surfaces, geological formation and remodeling of planetary structures based on their impact history, and how such impacts have shaped our Earth through both deep and recent time scales. But mostly he is one of the world's experts on blowing holes in things. Most people with expertise in this area talk about entrance and exit wounds in forensics, calculations of ballistic impact effects for launches and weapons design, or demolition issues that arise when blowing up buildings. Pete thinks bigger. He blew a hole in a comet to see what was inside it. He blew a huge hole in the moon to look for traces of water. Pete enjoys himself immensely when he gets to watch the ejecta fly, whether it's tons of rocks, ice, and volatiles from the comet Tempel 1 or a pile of toothpicks used to model the trees that were blown flat during the Tunguska event in Siberia in 1908.

I was invited along to help Pete with a new way of recording an impact. Using broad-spectrum recording techniques, we thought we could capture information about the impact and the material ejected from the target zone to create a sound-based 3-D model of how material flew away from an impact site. This could be important and useful for several reasons. First of all, the flow of ejecta after the impact is based not only on the physical characteristics of the projectile (our model asteroid) and target (a dish full of sand) but also on the way in which energy is transferred from one to the other. Think of a rock hitting water—those rippling circles that come from the center are waves of vibration caused by the transfer of energy from the rock to the water. The same thing happens when a rock hits something solid (like an asteroid smashing into the Earth), only it's harder to see.

The transfer of impact energy can be analyzed in part by the multiple pressure waves that emerge from the impact, both as a result of the initial impact itself and from the way the ejecta move through the surrounding medium (for instance, air or water). Sound is a great way to model wave phenomena such as impacts and pressure flow. So after Peter set up the sand target and the techs loaded the gun, I placed a series of seismic microphones on the platform around the target, positioned a couple of pressure zone microphones (PZMs) along the chamber walls, and mounted an ultrasonic microphone near the port that the projectile would come through. The spacing between the mikes let us know how fast the sound was traveling, which in turn would let us start building up our 3-D sound model. My wife started up the digital recorders, the gun barrel was raised to the appropriate angle, the chamber was depressurized until it was nearly a vacuum, and we were all sent into the data center, a room full of video monitors that would let us watch the impact remotely, through the high-speed cameras. (It would also keep us safe from any possible incidents involving meteoritic-speed projectiles, gunpowder, and flammable gases, though no such episodes have ever occurred in the gun's forty-year history.) In the darkened room, the firing alarm went off and all we heard was a muffled thump. But the video screen shortly showed a slow-motion view of the pile of sand lighting up like a thousand suns followed by a hollow cone of sand and glass (from fused sand particles) flying outward, some particles moving even faster than the 5 km/sec of the original projectile and turning the chamber into a maelstrom of glass and dust.

We went back into the room, where Peter opened the chamber and gave one of his huge happy grins at the new crater while I fussed around, checking if any of his "babies" had destroyed my microphones. I tend to be rough on equipment, but my level of abuse usually is limited to dropping a microphone in a pile of bat guano or putting the recorder's batteries in backward, not subjecting them to a miniature Martian sandstorm. Luckily, everything was still working, so the target was cleaned, the sand replaced, and another projectile loaded, but this time the chamber was left full of a normal Earth atmosphere to simulate a strike here rather than on some airless body in space. Once again we checked the recorders, then retreated to the data room and waited for the alarm. This time it was different. Again, through the safety doors and the battleship-thick armor of the chamber, all we heard was the thump of the ignition, but the presence of atmospheric gases changed things. Again there was the cone of ejecta, but there seemed to be a *lot* more motion of sand particles. And it wasn't until I actually listened to the sound recordings that I could tell the difference.

The sounds from the first experiment in the vacuum were

very discreet, almost dull. The seismic microphones (called geophones and used to record earthquakes) picked up the initial impact and some faint pattering sounds as some of the sand hit the walls of the chamber and fell back to the target plate. The PZM microphones were silent—no surprise there, since most microphones used for what we think of as normal recordings detect changes in air pressure, and the near vacuum of the chamber gave them little to work with. The ultrasonic microphone was similarly quiet.

But the impact experiment with the atmosphere was radically different. The ultrasonic microphone picked up a submillisecond *whiiisssssh* of the projectile's flight right before impact. But rather than a thump and some gentle pattering in the seismic microphone and silence from the in-air microphone, we heard an explosion of sound at the impact, followed by almost a full minute of a sandstorm that sounded like it could have buried the Sahara. The presence of an atmosphere had completely changed the ejection and acoustics dynamics of our simulated asteroid impact, changing it from a dull thump to a raging event of continuous noise. And I began to see how these contained simulations can help us understand events that happened long before the invention of recording equipment or even ears to listen with, around the time the Earth was born.

About four and half billion years ago, the Earth coalesced from a pile of dust and rubble into a large rocky sphere surrounded by a cloud of gases. It was a noisy place, with a constant bombardment by chunks of rock, the occasional comet, and asteroids trying desperately to figure out a place to orbit without running into cosmic traffic, along with the eruption of volcanoes that redeposited liquid rock and hurled boulders everywhere. A few hundred million years after this, another planet, about the size of Mars, smacked the new Earth a glancing blow, ripping off chunks of the crust and sending the debris into orbit, where it slowly recoalesced into the moon.

It must have made a hell of a bang.

The impact probably also took a lot of the atmosphere away with it, which would have quieted things down a bit on anything but a seismic level. But the bombardments continued, each impactor bringing not just more rock and metal but a new load of volatile ices and frozen gases, creating a new atmosphere, delivering the water needed to cool the Earth, and bringing a new sound—the sound of rain, condensing out of the massive amounts of water vapor in the second atmosphere and forming the Earth's seas.

The Earth was growing loud again, with the sounds of water added to the volcanic purges and exploding meteorites. The sounds spread not only as high-speed shear waves through newly cooled rock, but also as spherical rumbles through the atmosphere and cylindrically spreading growls through the new oceans. The Earth created its own soundtrack, albeit one formed almost exclusively from what we would call noise, an almost equal (or at least very broad) distribution of energy across the entire acoustic range.

A few hundred million years later, during what is called the Late Heavy Bombardment, the soundtrack surged again as Earth (and everywhere else in the solar system) became a target for a rain of yet more asteroids. But during this time, an odd thing happened: somewhere a small subset of the churning organic chemicals in the oceans (or deep beneath them, near hydrothermal vents, according to some theories) started to replicate themselves. Amidst all this noise and vibration, life was born. The earliest life-forms for which we have any fossil evidence—the great green-blue algae stromatolite mats, billions of years from anything that could actively listen—were churned about by waves and vibrations in the primordial seas, exposing fresh surface area to methane to create the oxygen in the atmosphere we need to breathe. But as these early forms quietly poisoned themselves almost out of existence by oxygenating the atmosphere over the next 2 billion years, they set the stage for their successors, the eukaryotes.

The earliest non-autotrophic eukaryotic life-forms—meaning those that didn't just sit and photosynthesize—started experimenting with the externalization of proteins similar to those that made up their internal cytoskeleton. These proteins formed self-assembling chains that made small mobile hairs, known as cilia, that let these organisms move around. And in moving around, they found more food. This change transformed life-forms from passive to active players in the nascent ecology, allowing the evolution of the first predators.

Soon a variant on these cilia emerged with a different function. Rather than moving like oarsmen to propel single-celled animals, this new type of cilia, called primary cilia, served to open or close small channels in the cell membrane from which they projected. When the cilia were bent in one direction, they would open the channel; bending them in the other direction would close the channel. This changed them into sensors, detecting motion in the fluids around them. The organisms that first developed these were simple, but they had taken the first real leap in the sensory world. They were the first to sense and use vibration as a means of detecting changes in their environment—not just sensing changes in the fluid flow that might help them move from a food-sparse location to a food-rich one, but actively picking up on motion farther away that might indicate prey. Vibration sensitivity was one of the first telesensory systems, one able to detect changes in the environment at a distance rather than directly on or adjacent to the cell surface.

It would be tempting to say that these cilia are the ancestors of the tiny hairs that detect vibration in our ears today. But evolution is messier than that. Vibration sensitivity is indeed based on the displacement of hair-like cilia, but the evolutionary road to the hair cells that fill the modern vertebrate inner ear does not begin at the earliest flagella that spun single cells around the Archaean seas. Instead it starts with the emergence of mechanosensory neurons in multicellular organisms, probably similar to ancestral jellyfish, about a billion and a half years ago. It was something on the order of another 400–500 million years before what looked like modern sensory hair cells would emerge, and another billion years or so before these hair cells organized into dedicated sensory organs to detect motions in fluid in our early vertebrate ancestors—basic inner ears.

Take a second and think about what an ear is: an organ that senses the changes in pressure of molecules. We tend to imagine ears hearing music or car horns, but what they are really noticing is vibration. Early vertebrates used vibration sensitivity for two different purposes. One was to monitor changes in fluid flow right around their bodies, using what is called a *lateral line* system, still found in almost all fish and larval amphibians still around today. The second was used to monitor shifts in *internal* fluid flow in specialized organs located on each side of their heads. These structures had no specialized organs for picking up airborne sounds, since back then everyone still lived in the seas, but they were used to detect angular and linear acceleration of the animal's head. These organs, called the *semicircular canals* and *otolith* organs, respectively, were internal vibration sensors that measured the acceleration of the animal's head as it moved. Even the earliest vertebrate fossils with inner ears (*Sibyrhynchus denisoni*, a particularly weird-looking relative of the shark) show these structures. These organs formed the basis of the vestibular system—an acceleration-sensing system tightly synchronized with most other senses and the musculoskeletal system that lets the animal move in a coordinated fashion and fight against the pull of gravity. But for *Sibyrhynchus denisoni* and its fellow early vertebrates, it was also the beginning of hearing and listening. The saccule, the otolith organ that sensed the direction of gravity, would also vibrate in responses to pressure changes in the water. In other words, the ear had come into existence, and living things began listening.

Hearing in these early vertebrates was probably relatively limited compared to many of the examples that are around today—after all, we've had over 350 million years to play with variations on the theme. Descendants of *S. denisoni*, contemporary sharks, have relatively high auditory thresholds, that is, the sound has to be pretty loud for them to respond to it. They also have very limited ability to localize sounds underwater. This is endemic to listening under water—the speed of sound in water is about five times its speed in air, due to water's greater density, and so it is difficult to figure out where the sound is coming from based on differences in the sound on one side of the head compared to the other side. But they had other senses to help out—not only vision (which is relatively limited in sharks as well), smell, and electrosensation (which allows contemporary sharks to pick up on the neuromuscular responses of prey swimming nearby), but their lateral line system as well. All these senses were coordinated to form a sensory world. And right there is the first watershed that separates the world of sound before life developed from what it became after life emerged: the need to map perceptions onto the brain from the senses measuring vibration, photons, acceleration, and chemicals.

What we think of as sound is split between two factors, physics and psychology. The physics comes into play when trying to describe the parameters of a sound—its frequency (how many times the medium vibrates per second), its amplitude (the difference between the highest and lowest pressure peaks in a given vibration), and its phase (the relative point in time since the vibration began). In theory, if you could completely characterize just these three factors, you could completely describe a sound. That may sound simple enough, but outside of an acoustics laboratory, sound is much more complicated. It's pretty rare to find an isolated sound generator attached to a calibrated amplifier and speaker dead ahead of you on your daily commute, and if you did, you would be more likely to call a bomb squad than use it as a useful environmental signal to help you cross the street. Acousticians often use these simple, very controlled sounds as the basis for describing what sound does and how it works, but using them to describe sound in the real world is sort of like asking a physicist to describe the motion of a herd of cows. The physicist can model the herd's behavior perfectly, with the proviso that these are spherical cows moving on a frictionless surface in a vacuum.

As with the cows, the real world is much messier acoustically, especially in the primordial seas where the first listeners were born. While the physical aspects of sounds are characterized by their frequency, amplitude, and timing or phase, sound in the real world changes radically based on the details of the environment, and the devil is in the details. Sounds in the environment are emitted by almost anything that interacts with anything else, and once emitted, are affected by almost everything in that environment, until the energy lost in those interactions attenuates the sound into background noise. With enough patience, equipment, and computational power (and a large enough budget), you could do a reasonably good job of modeling what would happen to that sound—and in fact this is the basis of a great deal of the recording industry's post-production work. It's also a large portion of what the vertebrate brain does, integrating all the physics based on sensory transduction and assembling it into a perceptual model of the world outside so that it can be acted upon. Behavioral action and its underlying causes are the basis of psychology. Physics goes on whether there is a listener or not— all trees falling in all forests make sounds, regardless of who is present. But once a listener (or viewer or smeller) enters the world, everything changes. The physics of the world is separate from the psychology at the level of sensors and neurons, which is why we need a new term to describe it—psychophysics.

We think that by observing the world around us, we are actually seeing or hearing or tasting or touching what is going on, but we are not. We are interpreting a representation of the world, created by remapping a form of energy into a usable signal. All sensory input—no matter which form of energy it uses—is remapped. The initial energy of any stimulus, such as sight, smell, or sound, causes some change in the receiver, which is then transduced into a different form and passed along as *sensation*. *Perception* is the integration of sensations into a coherent model of the changes in energy that surround us. An atom of perception is the remapping of a single event in the physical world from a single type of sensor.

When you add up all these individual percepts, what you get is the *umwelt*, the world we build from our senses. For example, color is the psychophysical remapping of the wavelength of light, with brightness a remapping of the amplitude of light (the number of photons received). Touch—whether light pressure or deep pressure—is a remapping of the mechanical distortion of a structure. Smell is the result of the binding of specific chemicals. Sounds are psychophysical remappings of vibratory signals— changes in pressure in some medium such as air, water, dirt, or rock.

By becoming a listener—a receiver of auditory information, rather than just being acted upon by surrounding vibrations, as our stromatolite ancestors were—an organism becomes an active participant in the transfer and detection of acoustic energy. Early organisms developed a need to map the energy of sound via biological transduction, to create nerve impulses that stood for the sounds, sometimes even mimicking them (as in cochlear microphonics), but never with a 1:1 map. The reliance on the temporally fuzzy biological system of the brain means that you have to remap the energy of sound onto its representation in the perception via sensation.

But psychophysics is a *personal* remapping of physics, one that changes across evolution and development. At an evolutionary or species level, it's why you can sit on your porch on a summer night and comment on how quiet the country is, while a dozen bats are flitting over you screaming at subway-train loudness levels while hunting bugs, but at frequencies you can't hear. It's why elephants in urban zoos don't do well when they are located near highways—because the low-frequency rumble of automobile traffic from a kilometer away interferes with their infrasonic communication. At a developmental or individual level, it's why teenagers are drawn to loud music as a stimulant, and why an elderly person with normal age-related hearing loss can seem paranoid as his or her ability to monitor the environment decreases. The birth of psychophysics was the first step in the emergence of mind.

Shortly after life became complex enough to need psychophysics, it started doing something interesting: it began contributing to the sounds around it. Before life, all sounds on Earth—crashes of waves, susurrations of the wind, crackles of lightning—were noisy in the sense that they provided a a constant flow of acoustic energy scattered almost randomly across the spectrum (with the exception of the occasional low moaning sounds of wind across hollow stones or the brief tone of singing sands blowing across dunes). But with the presence of increasingly complex multicellular life and the development of listeners, the sounds of the Earth began to change. Vibration sensitivity arose because of an evolutionary rule of thumb: whenever there is a niche rich in resources, something will emerge to fill it. Early vibration sensitivity arose as an early warning system telling of changes in water currents around simple organisms. Shifts in these local water currents could mean anything from a wave passing by to the approach of a predator or the presence of prey organisms nearby. But once organisms grew complex enough to actually listen to their environment, there was an entire sensory niche open for exploitation. Animals could *make* sounds. This was a leap in animal behavioral complexity. Unlike vision, which relies on passive detection of light energy, usually from sunlight, sound provides a whole new communication channel, one capable of operating in the dark, around corners, and without being dependent on line of sight.* Sound was suddenly not just an early warning system, but used as an active way of coordinating behaviors across long distances between and within species.

We don't know much about the first vocal animals aside from the fact that it is unlikely they were vertebrates. Invertebrates have always outnumbered vertebrates in both biomass and diversity, and they were here billions of years before we were. But there is little fossil or genetic data on the emergence of vocal behavior. Both ways of tracing elements of biological history require a common ground, a presumptive chain of commonality, or at least a limited number of adaptations to check out. There are hundreds if not thousands of adaptations that allow animals to make sounds. Modern snapping shrimp create deafening 100 dB underwater choruses by snapping an oversized claw so quickly that it pulls gas bubbles from the water; these bubbles then explode in a pressure wave so intense they make pulses of light. Spiny lobsters make a ratchety violin sound by rubbing an extension of their antenna across a raspy area on their shell near their eyes. But even vertebrates have evolved extraordinarily diverse ways of making sounds, many not needing complex vocal apparatus. Once animals started developing hearing, almost any generated sound could carry meaning.

The earliest vocal animals probably didn't use dedicated sound-generating structures, as almost any controllable sounds can be used to communicate, as shown by the winner of the

---

* This is not to demean the fabulous array of organisms that create their own light to communicate. For a wonderful overview, I recommend *Aglow in the Dark* by Vincent Pieribone, David F. Gruber, and Sylvia Nasar. It covers the history of bioluminescence in plants, fungi, and invertebrate and vertebrate animals, both natural and genetically modified.

award for most unusual vertebrate communication structure, the herring's fast repetitive tick (FRT). Herring have excellent hearing, although no one ever quite understood why, since they didn't seem to have any vocal apparatus and were never heard to make sounds. But a study in 2003 showed that large groups of herrings would release bubbles from their anuses that made ultrasonic noises that fit nicely into their hearing range, thus demonstrating communication and social coordination by herring FRTs (and guaranteeing Ben Wilson, lead author on the study, a place in the history books of acronym generation). But this somewhat scatological example makes an important point: most totally aquatic sound-making animals do not use what we think of as typical vocal mechanisms. Terrestrial vertebrates usually use modifications to their respiratory tracts to drive air over some tissue that can vibrate in a controllable fashion, whether it be an opera singer driving air from her lungs over her vocal folds and through her mouth or a leaf-nosed bat driving it through the incredibly baroque folds overlying its nose, forming the beam for its echolocation signals. But underwater vertebrates rarely use such systems, as the greater density of water means that pushing water through small openings to make vibrations takes a great more energy than does pushing air through the same structures. So many underwater animals, such as sea trout, will use stridulation, rubbing stiff fin-like structures against each other, similar to how terrestrial crickets sing. Others, like the toadfish, will rub sonic muscles against their air bladders, sort of like rubbing your hand across a balloon to make it squeak or hum.

And even though the earliest noisemakers were not contemporary sea trout, toadfish, or even farting herring, they probably used these and other mechanisms to make sounds. The more sounds they made, the more complex their behaviors could become and the farther their social and survival webs could reach, traveling on the backs of the waves of sound. And as more and more living things joined the chorus, something very interesting happened to the Earth: its sounds changed. From a place of percussive impacts and landslides, rasping sands, and the white noise of wind-driven storms, living things began to make purposeful, *harmonic* sounds of their own, more tonal, controlled, and packed with meaning. The math underlying the sounds became more integer-like, less random. The Earth's acoustics expanded from incidental noise to songs. And as the biosphere grew more complex, Earth developed an *acoustic ecology*, a measurable change in the vibrational energy of our land, sea, and air, driven by the emergence of life on our planet.

Even though the rain of iron and ice from the sky has diminished to the occasional fireball over Canada or piece of falling space junk, Earth is still a noisy place, but at least now there are listeners around to appreciate the sounds of a living planet. And we should. Every impact that has rung the Earth like a giant bell, every grinding earthquake formed by one tectonic plate sliding under another, every tidal wave, and every soft wind has contributed to the formation of life, shaking up the primordial ooze and adding energy to prebiotic chemicals, forcing them to collide, interact, and begin replicating. And to this day, the beat goes on.

What does sound do to us today? The Earth still surrounds us with its intrinsic sounds—earthquakes, wind, rockfalls, blowing rain and snow—but now a great deal of the sounds that fill the thin skin of our planet between the deepest oceans and the atmosphere come from living things. Life, as any new parent knows, is a very noisy thing. And sound and noise continue to drive the evolution and development of life, from shaping our environment down to forming our very synapses. By listening, we hear not only the sounds of individual animals but the songs of a healthy biosphere.

I was recently reading an article on attempts to clone a thylacine—a Tasmanian tiger (which is not a tiger but a dog-sized marsupial hunted to extinction in Australia in the last century). In search of what they must have sounded like, I found a few short films of wild thylacines, but they were all silent. Then something occurred to me: I will never hear the sound of the thylacine. With every species extinction, we lose something from the acoustic ecology. No more sounds of the wings of a flock of passenger pigeons a million strong. No awkward squawks of dodos. And despite Hollywood's best efforts, we will never hear the roars or songs of dinosaurs. But going into the field and just recording the surroundings (a common technique in bioacoustics, the study of biologically based sounds) often gives surprising results that lend hope for species that haven't been seen for decades and are feared extinct. The great Ivory-billed Woodpecker, North America's largest woodpecker, has been presumed extinct since 1944 due to habitat destruction. But based on recordings made in 1935 by Arthur Allen, scientists as recently as 2005 have claimed that they have heard if not seen this bird, leading to hope that a small population has survived in the southern United States. (And of course there are the innumerable recordings of Bigfoot stomping and growling in the Pacific Northwest, despite the fact that subsequent audio analyses have shown that these sounds easily could have been made by anything from bears to humans.)

As humans spread into undeveloped nature, we fill the spaces with human sounds, human voices, traffic, music, advertising jingles, street-corner preachers, and bands. Humans bring the noise. And our ability to consider increasingly loud and noisy environments as "normal" and "home" makes it worse. It takes a radical shifting of attention or a change of place to realize the value of quiet. Rather than enriching our space with the sounds of life, we are crowding out the calls of other species.

But on the other hand, human technology and innovation are letting us hear things we've never heard before. Whether it's ultrasonic microphones that let us hear the bar-brawl loudness of a swarm of bats screeching at 120 dB above our heads on an otherwise quiet summer night or interplanetary probes letting us listen in on the sounds of the thunder on Venus or the winds on Saturn's moon Titan, our technology allows us to hear more and further than any other species on this planet. But to appreciate the sensory richness in which we are embedded, we need to be quiet once in a while when we are in a new place, and just listen.

Chapter 2

SPACES AND PLACES:
A WALK IN THE PARK

W HEN I WAS a kid, my mother used to listen to a radio personality named Jean Shepherd. He was mostly known as a storyteller and a humorist (and is primarily remembered today for the film version of his tale "A Christmas Story"). But I remember him because he used to tell stories about how things sounded. He used to say that when he went traveling, he didn't bring a camera; while everyone took pictures, he recorded the sounds of the places he went.

I particularly remember his shows about the 1964 World's Fair, when he walked around recording the sounds of the different pavilions. At the time I was a constant visitor to the fair because I happened to live about two blocks from it, and I remember being excited listening to these shows because I knew *exactly* where he was talking from. I heard the songs from the Small World exhibit or the sounds of the extruding machines that made the plastic mold-a-rama dinosaur toys they sold at the Sinclair Oil Dinoland.

On one episode I liked, he started out by talking about a vendor named Ernie who sold popcorn in front of the press box at Comiskey Park and was just as famous as some of the ball players. This was not a subject of particular interest to me, until he started imitating Ernie's famous call and telling how it would echo all around the park, bouncing off the left-field wall and the scoreboard in right field, then echoing back through the "great cavern of the stands, just floating through the whole park, just part of the rich effluvia of life." From there he led into talking about listening to the world from four thousand feet above the ground in a hot-air balloon. He talked about how he could hear dogs barking, conversations, kids rattling sticks on fences—things he never could have heard at ground level or from an airplane, where you're enclosed and surrounded by throbbing engines. And he wondered why sound carried so well up there, saying that maybe an acoustician or a meteorologist could tell you but he didn't know.

Decades later, I still remember that particular show, and realized that I could now answer his question.* Sound waves propagate through the air in a spherical pattern, widening out from whatever made the sound. In theory it could spread out nearly forever, losing energy based on the square of the distance from the sound source, but things get in the way, making the sound lose energy. Sound can be refracted, bent by things as simple as changes in the density of the air; reflected, bouncing off hard surfaces; or absorbed into a surface, adding a bit of heat

---

* Actually, my aging memory was greatly helped by one of the more wondrous resources for someone seeking sound recordings: the Internet Archive, run by an old schoolmate of mine, Brewster Kahle. Brewster's mission is to archive all the information in all the media in the world, and he has a pretty good start on it. If you want to hear this particular show, it can be found here: www.archive.org/download/JeanShepherd1965Pt1/1965_03_24_Pop_Art_Worlds_Fair.mp3.

to the structure but losing the energy of the sound. As things refract, reflect, or absorb sound, the sound gets distorted, losing strength and cohesion. The more things that are in the way, the more the sound fades away. But the part of the sound that rises above the clutter on the ground is unobstructed, essentially giving listeners above the ground (and not surrounded by throbbing aircraft engines) a natural hearing aid. Just changing the place you listen from can radically affect what you hear, be it four thousand feet in the air, under the ground in a bat cave, or just in my office after I move a bookshelf from one place to another.

Aside from an appreciation of the way sound changes in different spaces and places, I picked up something else from his tales. When my wife and I travel, we usually have a camera stored somewhere, but more often we pack enough audio gear to make passage through the TSA checkpoint interesting. The most interesting time was when we were going off to record the Eiffel Tower—my wife is an artist whose work has a heavy focus on sound, and she wanted to record the actual sounds of the tower, including the low-frequency infrasound normally undetectable by humans. So after she got the requisite permission from the Société Nouvelle d'Exploitation de la Tour Eiffel (SNTE), we packed up four digital recorders, some in-ear binaural microphones, several hundred meters of cable, and eight geophones—seismic microphones normally used to record earthquakes and drilling operations. You know, just basic tourist stuff.

These were "can"-type geophones, relatively small but heavy brass cylinders with electrical leads on the ends. In other words, they look like pipe bombs. On 100-meter leads. Plus a lot of electronic gear with blinking lights and timers attached to them. Surprisingly, the security agent at Kennedy Airport in New York simply asked what they were and then gushed about how much she loved Paris, wished us luck, asked where she could hear the recordings, and let us go. Then, on our second day in Paris, after I'd cleared everything with the local SNTE office and began duct-taping these geophones to the south pillar of the Eiffel Tower (while my French-speaking wife was off with the video crew), I heard some rather frantic voices yelling something I didn't understand. Upon looking down, I saw a number of gendarmes directly under me, armed with machine guns and clearly insisting that I either do something or stop doing something. Waving my letter from the head of security (who had not informed the local gendarmerie) and saying "le papier!" repeatedly with a New York accent didn't help much, but my wife managed to return and calm things down before I got shot in the name of acoustic research.

But it was worth it. We were allowed to record the tower's ambient human-audible sounds and infrasonic vibrations from the base, from the apex, from the underground mechanical room (which was strictly off-limits at the time), and from the escape chute at the top of the tower under the radio antennas. If you go there and listen, the sounds you usually hear are those of the voices of thousands of visitors in their hundreds of different languages, the horns of the taxicabs whose drivers are blocking traffic in yet another strike, the two-toned European sirens as the police try to get traffic moving, and other sounds of the urban Parisian environment, ranging from the weather to the cooing and flapping of the omnipresent pigeons. But standing on the topmost platform, separate from most of the other visitors, I confirmed what I think of as the Jean Shepherd effect: when the wind died down I could occasionally hear conversations from individuals almost a thousand feet below, their voices (in French) rising undistorted along a specific clear radian.

But the tower has an unheard voice of its own. You may think that 7,300 tons of metal, 2,500 tons of stone, and 50 tons of paint would keep the tower rock steady, but the massive structure vibrates constantly at subsonic frequencies. Gustave Eiffel claimed that his design for the structure was intended to minimize wind resistance, and his success is demonstrated by the fact that even in strong winds the tower only sways 2 to 3 inches (as opposed to the 10 to 12 feet the World Trade Center used to sway). But the tower responds to and propagates all sorts of low-frequency vibrations—the pounding of visitors' feet, the motion of the giant chariots hauling the counterweights that lift the nineteenth-century elevators (still using most of the same technology that Gustave Eiffel installed), the wind shaking the antennas and lighting system.

These sounds are not heard by humans and are the unheard song of the Eiffel Tower. In large part this is because humans hear airborne sounds. Sound travels differently depending on the density of the medium. While the rigidity of the tower's structural members damps higher-frequency sounds, the iron's high density allows sound to move fifteen times faster than in air, which also means any vibration would travel fifteen times farther than it would in air, reflecting and reverberating within the entire structure. So any hard tap would essentially "ring" the entire tower briefly before being swallowed up in the general hum of other vibrations. But it was only by using seismic microphones that recorded from about 0.1 Hz to 20 Hz, the acoustic realm of earthquakes and landslides, and pitch-shifting those sounds back into our auditory range, that we were able to hear the tower breathe and moan, shudder and sway as if it were a living organism, reacting to the things that crawled on it and the winds and rain that blew through it.

The Eiffel Tower may seem like a special case, and it is certainly one of the most interesting sonic spaces I've ever visited, but all spaces and places have acoustic lives of their own based on their shape, their construction materials, what they are filled with, and, mostly, what sources of sound and vibration they are near. While most auditory research takes place in clean, well-lit laboratories with soundproof booths and nicely calibrated equipment that would make an audiophile drool, the rest of the world, where we do all our hearing, is filled with complex sounds that are composed of more than one frequency and vary in amplitude and phase over time scales both short and long. The range of frequencies that most humans can detect runs from the very deep bass of 20 Hz up to the screechingly tinny high end of 20,000 Hz (or 20 kHz). Each frequency has its own wavelength (the length of single complete cycle of a sound). This has important implications for how sound will change in a space. If we know how fast sound travels in a given medium, it's easy to calculate the distance one complete wave or cycle takes up. The very low frequency of a foghorn at 100 Hz has a wavelength of 3.4 meters (about 11 feet), whereas the ultra-high biosonar of a bat at 100 kHz has a wavelength of only about one-third of a centimeter (an eighth of an inch).

Here's a question: why are foghorn sounds so low-pitched? Consider the logistics. You have to get a signal from shore to something really far away, whose position is unknown and which probably is not visible, to prevent an unpleasant intersection of ship and rocks. You need a sound to accomplish this (though we also have lighthouses, of course), but why low frequencies? Because the lower the frequency, the longer the wavelength; if your wavelength is very long, the pressure changes will not be very affected by small things in its path. All the energy that goes into a single cycle is extended over a distance larger than any- thing likely to be in the way, and so low-frequency sounds can travel farther. A higher-frequency sound, such as a bat's bioso- nar, will have a much shorter wavelength and be more likely to bounce off, be absorbed by, or be otherwise distorted by smaller objects, so it won't travel as far. (Making lots of high-frequency sounds lets bats get echoes from very small objects relatively close to them, then proceed to eat them).

But echoes are not just for bats. My first remembered intro- duction to echoes was when I was a young child hunting for fossils in the Catskills while my parents were busy doing unin- teresting grown-up things. After I'd dug messily through enough dirt to uncover several tyrannosaurs, my mother, down the hill and across from a large stone outcropping about a quarter mile mile away, began calling my name, and I heard her voice re- peat, getting quieter and quieter with each repetition. My early wonder at the miracle of acoustic reflection was spoiled when she yelled, "Get down here now!" with only the emphasized "now" repeating itself. I hastened to obey, since clearly my mother had mastery of hidden powers; after all, she'd made the rocks repeat her messages. But in fact it was just her facing toward the very dense, relatively flat, and hence acoustically reflective rock wall that reflected her voice in a series of quieter and progressively more distorted versions. The sound traveled about a quarter of a mile away and back (about 2,600 feet), so at the speed of sound it took about one-third of a second for the first echo to reach me, losing energy in the higher frequencies, then hitting the slope of the rocky grade behind me and bouncing the sound both upward and back to the distant rock wall. This both decreased the intensity of the subsequent echoes because of scattering loss and attenuated the higher frequencies even more, until the directly heard "now!" overwhelmed the last of the reflections as they succumbed to the ambient noise even in the quiet mountain area.

But echoes are only one part of how objects change sounds. If you are reading this book in your bedroom, a train, a bus, or a classroom, every surface in that space—tables, walls, even other people—can and will change any sounds that are generated within earshot. Any surface that sound can strike will change it in some way, and the materials that make up or cover those surfaces will change the sound in a unique fashion. Even the simplest sound played in an uncluttered room, ten feet in any dimension with relatively bare walls, will generate thousands of overlapping echoes. And since each frequency has its own wavelength and hence changes in phase and amplitude as it strikes the surfaces, these echoes interfere with each other, sometimes making certain frequencies louder via constructive interference or quieter by destructive interference.

The summation of all these complex changes to the original sound is called *reverberation* and involves not only the tens of thousands of individual echoes but also the damping and amplification caused by constructive and destructive interference from anything in the space. Hard tiled walls and floors (such as in a bathroom, whose acoustic qualities may mask some small irregularities in your otherwise perfect shower-singing voice) are highly reflective, making a room rich in reverb but potentially sonically muddy. If the floor is carpeted or a ceiling has soft, irregularly surfaced acoustic tiles, the energy from the vibrating air molecules is more likely to be absorbed than reflected. This is the basis of sound attenuation in places such as offices; in locations where sound quality is critical, such as recording studios or anechoic rooms, people will often go one step further and place geometrically patterned soft foam on the walls to further increase sound absorption. You can see a simple example of this at home. If your stereo is in a room with hard floors, try placing a rug just past the speakers; you will be able to hear a muffling or damping of the sound as you lose not only the sound bouncing off that part of the floor but also the sound from all other areas the floor would have reflected it toward. Then try moving chairs in front of the speakers. You should hear a decrease in the higher-frequency sounds, as the lower-frequency sounds bend successfully around the chairs but the higher-frequency ones bounce off them. This gives you an idea of the acoustic signature of a space and some idea why proper speaker design and placement is in fact an art. Getting the best sound requires not just the best sound reproduction but an understanding of how acoustic energy flows toward the listener's ears.

Despite all the complexity, we are really good at figuring out information about a space based on these complicated acoustic signatures. But we are really bad at figuring out simple things such as distance based on the delays from echoes. I co-taught a course on psychoacoustics for a number of years and was always amazed at the results from a demo I did in class. I play two types of recorded sounds, a single musical note from a piano and then a scale followed by a chord. The first time I play them, I add simple echoes, with different delay lengths between the sound and the echo as well as a reduction in loudness based on normal losses from spherical spreading. The second time around, I modify the sounds algorithmically to create complex simulations of spaces of different size and contents. I then ask the students to estimate the simulated distance from the piano to the reflecting wall, as well as the size and contents of the simulated space.

Now, figuring out the distance represented by a single echo should be easy, because all you really need to know is the duration of the interval between the initial sound and its echo; the rest is very easy math. It's a bit like counting the seconds between a visible flash of lightning (moving at about 186,000 miles per second) and the subsequent sound of thunder (moving at the relatively poky one-fifth of a mile per second). Then, as your parents likely told you, you just divide by five to tell you how many miles away the lightning was. While this was probably done to allay your fears of lightning and thunder, this simple rule of thumb not only introduced you to the world of physics but taught you how to do consciously what your ears and brain already do automatically—figure distance from danger.

So, getting back to the audio demo, if sound travels at 1,000 feet per second and a sound has to travel to a reflecting surface and back again, you should be able to estimate the distance to the reflecting wall pretty simply. On the other hand, figuring out the parameters for spaces ranging from wooden coffins to Bryce Canyon just using reverberation ought to be horrendously difficult. And yet invariably, year after year, my students could barely tell which echo is from a more distant surface than the preceding one but usually were about 80 percent right in identifying the size, shape, materials, and contents of the various simulated spaces. These findings seem counterintuitive; figuring out the distance to the echoing surface is simple arithmetic, whereas computing the contents of an unknown space and even being able to figure out if there are bodies in the metal chairs or if the room is empty (something more than half of the students get right) requires millions of calculations of time and frequency characteristics so detailed that they would likely tie up a small network of computers for a significant amount of time. But again, your ears and brain, exposed on a daily basis to this complex acoustic world, carry out those computations and pass that information to your conscious mind in only a fraction of a second of listening. (While we humans are rather miserable at echo distance calculations, bats probably find the task trivial, since if they don't get it right quickly, they don't eat and tend to fly into trees—or, worse, researchers carrying nets).

Our ability to determine what a place sounds like often figures into the field called architectural acoustics—designing places to sound a specific way. Buildings, particularly ones with large open spaces such as airports and cafeterias, often are loud and so reverberant that the noise level makes it hard to even hear someone right in front of you. For example, the Main Concourse of Grand Central Station in New York City is a beautiful structure with enormous open spaces. The vaulted ceilings are 40 feet high and the walls are faced in granite, marble, and limestone. The result is that the Main Concourse becomes a giant very low-fi acoustic reflector for the sounds people generate and for the low-frequency rumble of the trains and traffic in surrounding areas (which are propagated at high speed along the steel and stone infrastructure). This makes the annual holiday concerts held there, not to mention announcements of train arrivals and departures, very difficult to make out. But even in an acoustically messy place such as this, there are sweet spots. If you descend one level into the low ceramic-lined arches across from the Oyster Bar restaurant and whisper facing the wall, you can be heard amazingly clearly on the other side of the passageway. This is because the hard, curved surface acts as a waveguide for quiet sounds directed along the surface, rather than scattering those sounds. Hence the name "whisper arches."

But turn-of-the-century train terminals were not typically designed around their sonic properties. Architectural acoustics is a huge field representing a multi-billion-dollar industry that is seen as increasingly critical for quality of life, particularly in urban settings. At the most basic level, offices, hotels, and even home builders spend a total of hundreds of millions of dollars a year on trying to limit internal noise pollution. Every concert hall, theater, or auditorium—anyplace that is home to something important to listen to—has most likely been designed to maximize the sounds you should hear, damp others, and minimize distortion (or has suffered from a lack of such design). The field of architectural acoustics is not a new one—the ancient Greek amphitheater at Epidaurus, from the fourth century BCE, was a marvel of acoustic engineering. Even though it was an open-air structure with no sound system, it is still legendary for the quality of its acoustics, and its acoustical qualities were never duplicated in its time. The underlying basis for its success became understood only in 2007 when Nico Declercq of Georgia Tech and engineer Cindy Dekeyser showed that the physical shape of the seats acted like an acoustic filter, holding back low-frequency sounds and allowing the higher-frequency sounds to propagate. In addition, the use of limestone, a porous stone, absorbed much of the random locally generated noise. (The widespread use of limestone in buildings in Venice is, along with the lack of vehicular traffic, a main reason the city is so quiet.)

More recently, eighteenth- and nineteenth-century concert

halls, with their huge open spaces, hanging draperies, and baroque ornamentation, often were carefully designed to maximize the flow of sound from the stage to every region of the audience. The varying sizes of the theaters' elaborate decorations created a rich acoustic environment in which the reverberations added to the sound rather than muddying it up. But with the adoption of cleaner geometrical lines in twentieth-century architecture, many spaces started to suffer from acoustic dead spots and muddy sound. I've heard stories from friends of going to concerts in New York's Philharmonic Hall (later called Avery Fisher Hall) and being unable to hear anything but the woodwinds or the strings. It took a complete structural renovation under the guidance of the late Cyril Harris, a master acoustical engineer, to be able to overcome the limitations of the original architectural cleanliness and oddly oriented seats to make the hall worthy of the type of music played there.

Careful listening for even a short period of time not only reveals the wealth of sound sources that surround us but can give us profound insights into both how space shapes sound and how the brain handles the remapping of sound into a psychophysical representation of the space we occupy.

The walk I'll describe here took place about 2:30 P.M. on a Sunday afternoon in late October several years ago, a clear day with a temperature of about 60 degrees, with very light winds. My wife and I took a walk through New York's Central Park; for you non–New Yorkers, Central Park is closed to most vehicular traffic on the weekends and hence is full of people riding bikes, skating, running, playing musical instruments, or trying to pretend they are wandering through some pastoral utopia complete with hot-dog vendors. We recorded the walk as part of a project of hers in urban bioacoustics. My wife was wearing in-ear binaural microphones attached to a portable digital sound recorder. In-ear binaurals are tiny specialized microphones that cover the human auditory range pretty well, from about 40 to 17,000 Hz, and fit into the ear canal, pointing outward. What makes recordings from these microphones unique is that since they sit inside the ears, they pick up sounds exactly the way your ears do, using the individual shape of your ears to create subtle changes in the waveform and spectra of sounds entering your ears. They also are wonderful for capturing subtle differences between the sounds that enter each ear individually. To set the stage, my wife and I, who are five foot six and five foot eight, respectively, were walking northbound along the asphalt road just west of the Metropolitan Museum of Art, with her on the right, closer to the roadway, and me on the left nearer the grass. You'll soon see the reason I provide this level of detail.

Although we walked for about an hour, we're only going to examine about 34 seconds' worth of the recording. Figure 1 shows two graphs, each divided into two channels, the left channel (from the left ear) at the top and the right channel (from the right ear) on the bottom. The graph at the top is called an *oscillogram* and shows the changes in sound pressure measured in decibels (dB) on the vertical axis, over time in seconds on the horizontal axis. The oscillogram is a standard tool for analyzing the change in strength of the overall signal over time, as well as identifying changes in the fine structure of a signal's waveform when a higher resolution (and thus shorter time sample) is used. Because this particular example covers about 34 seconds, it does not show the individual changes in the waveform that make up its fine structure, but only shows relatively gross changes in amplitude, called the sound's *envelope*. The envelope alternates above and below the center line (marked with - ∞). This particular recording was set so that the amplitude range went from 0 to 90 dB SPL, with 90 dB being approximately the loudest sound that could be recorded and any sound under 0 dB being too quiet for the microphone to pick up.* The center line (the *zero crossings* of the waveforms) is listed as - ∞ because it represents a complete lack of acoustic energy, a situation you are not going to run into without a lot of liquid hydrogen to cool things down to near absolute zero.

The lower graph is called a *spectrogram*, and it shows changes in the amplitude of specific frequencies over time. Even though it is derived from the same sounds as the oscillogram, it shows different details about the sounds. The horizontal axis shows time, aligned with the oscillogram above, going from the start at the left to about 34 seconds at the right. The vertical axis is labeled with different frequencies, starting here at 0 and going up to about 9,000 Hz. While the recording went up to 17,000 Hz, this particular spectrogram has been truncated since most of the sound energy of interest was lower than this point, and it also makes it easier to see details. In the graph, black indicates no acoustic energy, so every lighter colored point indicates a place where that particular frequency was detectable, and the higher the amplitude, the lighter the color. In listener's terms, a white horizontal line halfway up the graph meant that there was a loud tone of 4,500 Hz playing for the duration of the length of the line. Since the spectrogram shows frequencies discretely, rather than lumping them all together into a measure of the

---

* The minus sign (−) before the numbers is based on a convention for describing amplitude based on 0 dB (top) being the loudest sound you hear, with progressively quieter sounds as the peaks get closer to the center line.

Oscillogram—shows amplitude changes in pressure over time

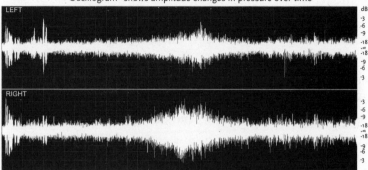

Spectrogram—shows amplitude changes of frequency over time

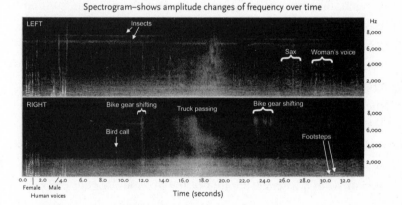

amplitude of all frequencies at a given time (as the oscillogram does), it lets you both detect individual acoustic events and identify their frequency content more easily.

Let's start our auditory walk with a couple of basic features about the soundscape. You'll notice that the gross envelope of the sounds on the oscillogram never seems to drop lower than about -15 dB. Compare that to the gray band on both channels of the spectrogram (slightly denser on the right), which covers the entire recording from about 2,000 Hz down toward 0. The band is almost even in density in the spectrogram, meaning that the acoustic energy is about equally distributed up to 2,000 Hz. This is a *noise* band. If we were recording in a quiet wooded or grassy area, far removed from a city, both the level and frequency would drop but not disappear. Acoustic background noise is present in any environment, and hence you almost never pay attention to it, except to subconsciously raise your voice a bit to talk or turn up the volume on your music. This background noise is based on thousands of blended sources, including voices, wind, and traffic noises, all at a distance, but contributing to a general, undifferentiable susurration. But one of the major contributors to this low-frequency noise band is ground traffic. The speed of sound varies with the density of the medium, and packed dirt, asphalt, and concrete are all dense enough to increase the speed of sound to up to ten times what it is in air. This means that sounds will travel through the ground up to ten times farther away before fading away compared to air. While an individual footstep, moving car or bicycle tire, or dropped package would probably make a truly noticeable noise only within 125 feet or so of a listener, because of the ground's higher propagation speed, you are suddenly having all sounds from within almost a quarter-mile radius all adding together, bouncing off the top and bottom of the roadway, interfering with each other, reverberating along the entire continuum of the solid earth, and using the ground surface itself as a giant, low-fidelity speaker. This also explains why the noise density (the brightness on the spectrogram) is higher on the right channel than the left, as my wife's right ear faced the roadway and her left ear faced the grassy area.

One thing I've noted in my travels is that every city has its own background noise signature. The in-air sounds vary based on the number of people, animals, or other noise generators in the area at any given moment. What differentiates the background noise of one locale from another seems to be based on the materials and density of roadways. The sound of cars driving over concrete is quite different from the sound of them driving over asphalt, and even the different compositions of asphalt found in different places can make the ground-based background noise perceptibly different if you pay attention. As mentioned earlier, Venice, Italy, despite being filled with thousands of tourists, is one of the quietest cities, due both to a lack of car traffic and to the common use in building facades of porous limestone, whose structure acts as a natural acoustic damper.

Let's examine some more specific features visible in the figure. If you look at the oscillogram, for the first four seconds there are several loud peaks, starting with approximately equal peaks in the left and right channels, followed by sounds that are louder on the left than the right. This is my wife's voice, followed by my voice. The higher amplitude of her voice is not because she is shouting, but rather because her voice is closest to the microphones in her ears, and also because the sound is not just traveling out of her mouth to the microphones in her ears but traveling internally as well, through her skull, using a sound pathway called *bone propagation*. The later peaks, starting at about 3.5 seconds, are from my voice, and are higher-amplitude on the left because I am speaking from her left side. If you look at the spectrogram, you will get more details that reveal a great deal about human speech acoustics. The spectrogram of her voice shows a series of bright white lines arranged vertically into bands that change slightly in position over time. These are harmonic bands, levels of similar frequency energy separated by null areas. The specific harmonic structure of these bands is defined by her vocal tract. She is saying, "We're behind the Met Museum," and the changes in the direction of the frequency bands indicate changes in the frequency of her voice as she switches between vowels and consonants. These changes in the frequency bands are called *formants* and identify the basic acoustics of human speech sounds. Notice that the bands in her voice (up to about 2 seconds) extend in a layer up to about 6,000 Hz on both sides, getting brighter in the lower frequencies, whereas those in my voice only go up to about 4,000 Hz and are brighter on the left channel (where her ear is facing me) than on the right, with almost nothing that appears to go above 3,000–4,000 Hz on her right. This is a demonstration of the sound shadow created by my wife's head, which limits sounds above about 4,000 Hz from wrapping around her head to the other ear. I am saying, "We're on Museum Drive North." If you look closely at the harmonic bands from my voice at around 4 seconds, you can see that not only does most of the energy from my voice stay lower than hers (except when saying the "D" and "th," speech sounds that have a lot of noise in them) but the bands are closer together. These two factors are what define my voice as having a lower pitch than hers. It's not just the top end of the frequency range but also the spacing between the harmonic bands that help your ears decide on the pitch of a sound.

Another clear feature that is present almost from the beginning and shows clear left/right differences are the two fairly bright lines at the top of the spectrogram at about 6,000 and 7,000 Hz. If you look closely, you can see that these bands are made of very tightly packed vertical white lines repeating almost continuously for the entire duration of the sound. These are insect sounds (probably from cicadas) sitting in the trees in the grassy area to the left as we walked. These are also a serious contributor to the constant background amplitudes in the oscillogram, but notice once again how limited the sounds seem in the right side, the ear facing away from their source, as their high frequencies are masked by my wife's head. The only time they really show up in the right-hand channel is from about 10 to 12 seconds, when we are passing a very tall tree that is the probable location of our singing bug.

As we continue our walk, the next obvious event is a small, almost vertical dash at about 9 seconds and about 3,000 Hz. This is a chirp from a small bird off to the right across the roadway. It barely makes an additional peak on the oscillogram, even though you can clearly differentiate it from the background noise with your ears and by looking at the spectrogram. However, there is a deeper story to this simple call. If you examined that part of the spectrum in greater detail, you'd spot another small blip in the spectrogram at about 1,500 Hz, making the 3,000 Hz signal a harmonic band. However, the lower band is in the region of the background noise and hence is almost invisible. This has some important implications for animal communication, as birds that live in the city have to contend with this background noise fairly constantly. It's been shown in several recent studies that birds living in urban environments have to shift their calls so that they are audible above background noise, which results in some interesting changes in behavior and stress levels for our avian neighbors (just as it does for us).

At about 12 seconds, both the oscillogram and the spectrogram show a brief increase in amplitude; the spectrogram reveals mostly high-frequency noise with faint bands in it. This is the sound of a bicycle changing gears as it passes us on the right in the roadway; the low-frequency (and relatively low-amplitude) sounds of the wheels on the asphalt are not detectable amid the background noise. But compare this to an event that starts at about 14 seconds in the oscillogram and continues until about 20 seconds. The sound builds up quickly, again particularly on the right side, and then peaks somewhat later on the left side. When you look at the spectrogram, you see a similar pattern, but the sound is quite dense all the way up to about 8,000 Hz, peaking on the right at about 17 seconds and the left at about 19 seconds. This sound came from a service truck driving slowly northbound on the asphalt. The difference between the left and right channels and in the shape of the spectrogram and oscillogram tells the complete story of this movement, with the sound detected first on the nearer side and peaking in amplitude as it passed the right-ear microphone, but partially blocked by the sound shadow of my wife's head until it is in front of her on the right, when there is a clear path for the higher-frequency noise to her left ear.

At about 19 seconds, the truck noise, a much wider band of noise at high amplitude, blocks or *masks* the insect sounds. This masking is not just a result of the fact that the truck is louder than the insect at this distance.* Since the truck puts out a much broader range of frequencies, the combined noise effectively blocks the narrow channel of sound that the insect uses for its song.

At about 24 seconds you can see another example of a bicycle passing, with several gear shifts, but at about 25 seconds you can see a slight increase in the amplitude of the oscillogram and a

---

* If they were the same distance from you, the cicada's 90 dB call would probably blow the truck's acoustic socks off. We're just used to having 30–40 feet between us and a calling cicada up in a tree.

series of alternating harmonic lines that are just observable in the low-frequency region and continue up to about 4,000 Hz, louder on the left than on the right. These harmonic bands are simpler than those observed in the human voices seen in the beginning and have a sort of staircase appearance. These are notes being played by a saxophone up on a slight hill to our left. The sax player had been audible for quite a while (starting at about 20 seconds), but once again it was not until we were closest to him that the higher-frequency harmonics appeared and the lower notes were evident in the band of background noise. You can compare these sounds easily to the harmonic bands from the woman's voice appearing a few seconds later as she approached from behind and to the left of us.

The last thing to notice starts at about 30 seconds and continues through the end of this recording: quiet spectral bands very far down in the background noise, alternating between about 200 and 500 Hz. Although at this scale they seem continuous, there are short gaps in the frequency bands, and the two bands in fact alternate. These are the footsteps of a runner approaching from behind and to our right, passing us at about 34 seconds. While it may seem odd that there would be such a difference between the left (lower-frequency) and right (higher-frequency) footfalls, this gives another interesting insight into acoustical recognition. If the runner had been running with perfectly symmetrical strides and feet perfectly aligned front to back, there would have been very little difference between them. However, I happened to notice as he passed us that the runner had a very distinct outward turn of his right foot and landed strongly on his left heel. The heel strike and full-foot roll off the left foot created less of a slap than the turned-out right foot (which had less surface area to work with and hence put more energy into a smaller footprint, generating a slightly louder sound at a higher frequency). Many years ago, there was a student in my lab who was interested in the question of whether you could identify someone just from the sound of his or her footsteps. She carried out a very neat little experiment where she had people of similar weights and heights wearing similar footwear walk and run down a hall while she recorded them. She then played these recordings back to listeners who had heard the subjects run previously. She found that people were remarkably good at identifying individuals just using these simple sounds—another example of your brain being able to carry out extremely complicated identification and analyses based on subtle acoustic cues. And if you think this is just an interesting academic exercise, bear it in mind the next time you hear footsteps behind you on a dark street, and realize that you too probably would not have to turn around to determine whether you are being chased by a stranger or by your roommate looking for the keys.

Chapter 3

LISTENERS OF THE LOW END:
FISH AND FROGS

JUST AS EVERY place has its own acoustic signature, every listener has its own plan for hearing what it needs to. There are about fifty thousand kinds of listeners in the vertebrate world, each with its own solution to the problem of what to listen to and usually very closely tied to the acoustics of its normal environment. Of all these, maybe one hundred have been explored scientifically (and most data are drawn from about a dozen, including zebrafish, goldfish, toadfish, bullfrogs, clawed toads, mice, rats, gerbils, cats, bats, dolphins, and humans).

At one level this is okay. Hearing in all vertebrates is based on using hair cells in some configuration to detect changes in pressure or particle motion and converting this into useful perceptions to help guide behavior. Once you get past the ears, vertebrate brains derive from a similar general plan—hindbrain receiving and sending much of the raw sensorimotor information, midbrain integrating both incoming and outgoing information, thalamus acting as a relay center to forward brain regions, and forebrain governing intentional behavior. But on the other hand, every species has developed its own solution to what it should hear. And to make it worse, every individual shows differences from that species's version of "normal," not only through genetic variation but also by what it has been exposed to over the course of its own life. So trying to understand hearing and all its variations from such a small sample of species can be very frustrating.

But people who study hearing ultimately want to understand their subject from a human perspective. We humans are primarily concerned with the human experience. Even though we are vastly outnumbered by all the other living things on Earth, humans are the ones who build the sound level meters, write and hopefully adhere to noise abatement laws, and have opposable thumbs to turn down the volume or switch the song. So we tend to view other species as systems whose features overlap human performance or interests. This limitation has become increasingly stringent in the last twenty years or so, as most research funding these days goes to "translational research"— studies that can be applied to human biomedical or technological applications. So we tend to focus on certain species that are currently perceived as being "useful." This is why you don't see too many papers on hearing in platypuses, lesser star-nosed moles, or giraffes (although these animals are beloved in electroreception, touch sensitivity and yawning research).* But the fact that humans share an evolutionary heritage with all vertebrates and the proven technological usefulness of biomimetics (copying nature's engineering) provide a lot of leeway in how we study hearing in animals. Even though we can't study all fifty thousand species of listeners, by looking at success stories—

* Giraffes are supposedly the only animals that do not yawn, according to Dr. Olivier Walusinski.

animals who have been around a long time and do some things very well—we not only learn about our own hearing but push the boundaries of what we can do with technology and bio-medicine in the near future. In this chapter we'll start in the shallow end of the pool with listeners at the low end of hearing, fish and frogs.

Life on Earth came from the sea, where life-forms took hundreds of millions of years to experiment with sticking their heads out of the water and risking the arid world of the air. Today most organisms still live underwater. Hearing underwater seems very complicated to us terrestrial types, as human ears have evolved to pick up pressure changes in sound in air, a rather low-density medium. These pressure changes vibrate our eardrum. The eardrum's vibration is then amplified by three small hearing bones or *ossicles*—the malleus, incus, and stapes (or hammer, anvil, and stirrup, for the Latin-challenged)—which in turn vibrate the oval window, the portal to the fluid-filled cochlea of the human inner ear where our hair cells convert these vibrations into usable signals.

But try sticking your head underwater while in a bath or swimming. One of the first things you notice is how odd everything sounds—the best description I ever heard was from a diver friend of mine who said, "Everything is simultaneously louder and softer and everywhere." Part of this is because human ears have evolved to translate vibrations in low-density air to the high-density fluid of the inner ear. Once you fill your ear canals with water, you've upset the system: your ear canal is now full of water, but your middle ear, housing the ossicles, is still full of air and hence passes along distorted signals.

Water is about eight times denser than air (similar to the density of your inner ear fluids and the rest of your body's tissues).

This difference in density means that water has a higher *imped-ance* than air—it takes more energy to start a sound underwater, but once it gets moving it travels about five times faster, confusing everything from our ability to identify sounds to our capacity to figure out where they're coming from. This is why divers without expensive communications gear rely on waterproof whiteboards or just banging on another diver's tank to get his or her attention. It also explains why no matter how loud you are playing your radio near the bathtub, once your head is underwater, the sounds in air just bounce off the surface due to the impedance mismatch, filling the bathroom with sound but leaving you to listen only to the slow plonking sound of drips from the faucet as they hit the water's surface.★

Yet fish have been hearing under water for hundreds of millions of years, despite the lack of any external or middle ears, plus an acoustic impedance almost exactly the same as the water surrounding them. Based on simple physics, the sound should basically pass right through them undetected. But certain adaptations have created enough of a difference to allow the vibrations to be captured by a fish's relatively simple inner ear. Fish pick up sound using the saccule, a hair-cell-laden sensory organ with an unusual structure. The saccule is oriented vertically in the inner ear, with hair cells that extend outward. The tips of the hair cells are embedded in a mucus-like mass that is full of dense crystals of calcium carbonate, like tiny little chips of bone. This structure, called an *otolith* (literally "ear stone"), is much denser

---

★ The only sounds that do make it through are caused by vibrations of the tub itself, unless you have suspended your radio directly over your bath with the speaker facing directly at the water surface. Then some sound gets through. But it would be easier to just take your head out of the bath.

than the surrounding tissue. When sound pressure waves strike the fish, most of the energy passes through its body, vibrating the fish along with the sound, but the dense mass of the otolith has a higher impedance than the rest of the tissue. The difference in motion between the fish and the otoliths on each side of its head bends the tips of the hair cells, changing their voltage and sending signals via the auditory nerve to code the characteristics of the sound. The best description I heard of this comes from Dick Fay of Loyola University, who explains that "in terrestrial vertebrates the animal holds still and the sound shakes the ear. In fish, the ear holds still and the sound shakes the fish."

There are many fish, such as sharks, rays, and skates, that get by just fine with this simple auditory arrangement. These cartilagenous fish tend to have relatively limited hearing, responding only to fairly loud sounds and a limited range of frequencies. But quite a few of the evolutionarily more modern bony fish have an adaptation called a swim bladder, an air-filled sac that is an evolutionary precursor to our lungs. The presence of a large pocket of air in the fish creates an impedance mismatch for sound, and many freshwater fish have evolved a modification to their vertebrae called Weberian ossicles that connect the swim bladder to the inner ear. These fish, which include goldfish, have quite good underwater hearing, with low thresholds for sounds up to about 4 kHz, and hence are considered "hearing specialists."* Clupeiform fish (herring, sardines, shad, and their relatives) take this one step further and have an extension of their swim bladder that projects into the skull and directly stimulates

---

* Goldfish are among the more beloved subjects for fish hearing studies, due to their ease of care and limited likelihood of eating someone who falls in their bowl, as opposed to the subjects of shark hearing research.

the inner ear, with work by Art Popper of the University of Maryland demonstrating that some of these fish can hear up into the ultrasonic range. So some fish have brought a bit of the atmosphere into their own bodies as a first step toward hearing the world beyond the sea. Which brings us to frogs.

I met my oldest friend, Greg, when we both jumped to catch a bullfrog at the age of eight (we landed on each other, the frog got away). When I was ten, Pablo the bullfrog (a temporary pet who did not get away from me in the pond but did manage to get out of his terrarium every night) cheerfully greeted my mother at the top of the stairs every night and serenaded her. Then there was Francesca, a subadult bullfrog whom I unsuccessfully tried to condition to turn her head to the left whenever I played B-flat on my synthesizer.

When I first applied to Brown University's graduate program, I went to the meet-and-greet for potential incoming students and met my soon-to-be graduate advisor, Andrea Simmons, and her husband (and my eventual postdoctoral advisor), Jim Simmons. Andrea at the time was specializing in how bullfrogs detect pitch. Jim was and is studying bat echolocation. After I'd been talking with them for a while, Jim said, "Frogs own the low end. Bats own the high end. Between the two you can figure out almost everything in hearing." I've spent about twenty years studying hearing in these (and a few other) species, and I still haven't figured out almost everything about hearing, but the point still sticks with me and drives my interests. It's led to everything from hauling a hundred pounds of recording gear into mosquito- and snapping-turtle-infested swamps to gene-screening injured frogs to try to identify the molecular basis for their ability to regrow their brains. I've had my hand stuck in the mouth of a male bullfrog intent on swallowing me whole, and had to hit the guano-covered floor of a bat-infested attic as nursing mother bats dive-bombed me with their babies hanging from their nipples. According to my doctor, I have developed the world's only recorded allergy to bullfrog urine.

One of the things that fascinated me about frogs is that, as amphibians, they are representative of some of the earliest forms that successfully ventured forth out of the water and onto the land. The fossil record shows that anatomically modern-looking frogs have been around for over 300 million years. This has created a perception that frogs are "simple" or primitive organisms and that by examining them we can learn only the basics of hearing. As an example of this, for many years, frog hearing was thought of as a simple mating call detector—that is, it was narrowly tuned to hear only the sounds of its fellow frogs. At first thought this seems to make sense—if your social behavior is dependent on calling and hearing other frogs of your species, why waste brain resources on extraneous noise? But frogs, like all animals, are not machines designed for a specific task but complex organisms in a complex ecology. To quote Dick Fay again: "The problem with hearing only your own species is that you'll probably get eaten by the first predator that makes noise outside your calling range." And as anyone who has tried to sneak up on a bunch of frogs knows, Fay is right—a chorus of bullfrogs filling the night with their low-pitched calls at a headache-inducing 100 dB will suddenly silence themselves with the first footstep within 20 meters. Their calls may include audible frequencies from 200 to 4,000 Hz, but they can detect ground-borne seismic vibrations orders of magnitude lower in pitch and amplitude than our ears can pick up.

But as with fish, you can't just lump frogs together. Frogs are a highly diverse order: some totally aquatic, some mostly terrestrial; some able to sit on the tip of your finger, others over 8 pounds and a foot long. One thing that does unite them is that in all known species, their social behavior and survival are dependent on their hearing. In fact, the presence of an obvious tympanic membrane (homologous with the human eardrum or tympanum) was one of the earlier ways that scientists used to differentiate frogs from toads, and its relative size compared to the eye is still how you tell the boy frogs from the girl frogs (males' eardrums get much larger than their eyes, whereas female frogs' ears are more petite). However, as with many things in the classification of animals by their external characteristics rather than their genetic relatedness, this turned out to be a problem, as some frogs have completely internal ears. This is one reason that *Xenopus laevis*, an aquatic frog, was initially called the African aquatic toad: being totally aquatic, they have evolved internal tympanic disks to allow them to hear each other calling in the muddy ponds that make up their natural habitat in Africa.

Xenopus laevis has been the darling of a lot of different types of research since the 1930s. Its eggs are very permissive structures that will express proteins transplanted from other species to create functional structures—for example, DNA that codes for pieces of cells such as neuronal ion channels will undergo translation in an *X. laevis* egg or oocyte, creating an egg with neuronal ion channels in it, allowing researchers to carry out precise studies of neuronal kinetics or pharmacodynamics not otherwise possible. Its tadpoles develop in transparent external eggs, allowing a great deal of developmental research to be carried out on them with greater ease than in a shell-covered chicken or uterus-enclosed mammal. The tadpoles themselves are also transparent, allowing the injection of dyes and tracers to determine cell fate mapping, tracing the course of development and migration of individually labeled cells even after they divide. And *X. laevis* was the first vertebrate species to be cloned. For quite a while it was the de rigueur amphibian for molecular and genetic studies, including hundreds of studies in how ears develop. However, with increasing knowledge comes the potential for the "oops" moment. While *Xenopus laevis* led the way for some of the most important work in developmental biology, it turns out that its genetics are very odd for a vertebrate—it is an *allotetraploid*, meaning it has four copies of each gene rather than two copies, as do humans and most other vertebrates, which are diploids. This means that *X. laevis* can never be selectively manipulated to delete or "knock out" genes or create specific mutations, and some of the genetic work done with the species is now in question. Much of this work that was done in this species is now being replicated in its closely related cousin *X. tropicalis*, a smaller, shorter-lived, but diploid species.

But *X. laevis* is an interesting animal for studying hearing. Although it is an amphibian, it is totally aquatic its whole life, from limbless swimming tadpole through four-legged carnivorous (and sometimes cannibalistic) adult. Like fish, it has a lateral line system, a series of external hair cells organized in interrupted lines called stitches across its head and sides to detect changes in water movement. Young *X. laevis* tadpoles use this system to determine the direction of the water current and orient themselves toward it to help maintain buoyancy and stability. This behavior, called *rheotaxis*, is used to help them maintain not only their position within a body of water but also their position relative to other tadpoles in their school and to detect sudden changes in water flow that might indicate the presence of a predator. Adults, which can get to be up to 10 inches long and weigh over a pound, typically lie near the bottom of a murky pond and so have limited access to light. The adult's lateral line is used to detect the motion of small insects or fish above them, which they then rush up and grab in their clawed fingers and shove into their wide spatulate mouths, their strange upward-looking eyes almost useless until they approach the water's surface. And like most totally aquatic animals, they have no external ears.

But the odd, flattened four-legged-fish appearance of these creatures hides the fact that they represent a major step in the evolution of hearing. While they share the fish's saccule (which may play a role in their hearing, particularly when they are tadpoles), they also have additional inner ear organs, called the amphibian and basilar papillae, small hair-cell-rich structures dedicated to hearing underwater. The amphibian papilla consists of a membrane stretching across the inner ear with hair cells that respond to lower-frequency sounds, from about 50 to 1,000 Hz, arranged in a loosely *tonotopic*, or frequency-specific, order. The basilar papilla is a smaller cup-shaped organ full of hair cells that respond to higher-frequency sounds, typically up to about 4,000 Hz. And while the saccule is still there, *X. laevis*, unlike fish, has a middle ear, consisting of internal cartilaginous tympanic disks, homologues of our eardrums, which are different enough in density from the surrounding tissue and water to allow pressure changes from sound to vibrate them and is connected to the inner ear by a piece of cartilage called a *stapedium*.

This sounds like *X. laevis* would be a peculiar species to use if we're trying to understand anything about humans. But strange and ancient as it is, *X. laevis* is an amazing model for acoustic social behavior, because it tells the story of sound and sex.

Xenopus laevis frogs live for love songs. Like all frogs, they depend on *phonotropism*—homing in on the calls of the opposite sex—to be able to find each other in the murky ponds they call home in the wilds of southern Africa (or in the somewhat less murky water of the lab aquarium). Unlike in most other species of frogs, the females do as much calling as the males, and it's the females who are in control. *Xenopus laevis* doesn't have complex singing apparati—it makes calls by using its laryngeal muscles to snap two cartilaginous disks together to create castanet-like clicks. It doesn't offer much in the way of tonal repertoire, but this type of signal doesn't require passing air over the vocal system and messing up the sound with bubbles, and clicks spread well through water, without distortion. Besides, as with all love songs, it's in the timing. Males produce a relatively wide range of calls, from slow amplectant calls—ticking away every few seconds—when they are in a loving embrace with a female to occasional chirps (especially when they are picked up) and growling click trains. But their most important call in mating is their *advertisement call*, a half-second-long sequence of clicks, slow at first and then followed by a rapid burst, repeated at rates of up to a hundred times a minute.

An advertisement call is exactly what it sounds like—it's a signal to try to attract females in the area and to warn off other males, and it is heard most often in response to a female's call. Female members of *X. laevis* have only two different songs—rapping and ticking, also made up of differently timed clicks—but they control the males' behavior. Ticking calls are quite slow, only about four clicks per second, and females sing this when they are not sexually receptive. A male hearing a ticked-off ticking female will often move away from her, for reasons that should be obvious. Rapping is a call females sing when they are sexually receptive; it is a series of clicks about three times faster than ticking, eleven to twelve clicks per second, that acts

like an acoustic aphrodisiac for any male in the area. Even playing a recording of a female *X. laevis* rapping song will make any
male in the area approach the source and try to mate with
whatever is making the sound. This often requires the lab tech
to pry it off an underwater speaker and do extensive cleanup
afterward.

While in human singing, the ability of the singer is based on
a great many physiological, cognitive, and behavioral factors
(especially practice, or else Auto-Tune), in frogs the males' and
females' songs are based more on physiological hardwiring. One
of my favorite lab experiments of all time was called the *vox in
vitro* or "song in a dish" by Darcy Kelley and Martha Tobias.
Tobias and Kelley removed the larynx from male and female
frogs along with part of the laryngeal nerve. When they stimu-
lated the nerve at the appropriate rates, they found that they
could actually make the disembodied larynx create sexually specific songs without the rest of the frog, but that even by changing the stimulation rate, they could not make an adult female
larynx call quite as fast as a male's larynx. This is due to sexual
differentiation in the type of muscle fibers in the larynx. Males'
laryngeal fibers are *fast-twitch* or Type II muscle fibers. This type
of fiber is metabolically suited for high-speed but relatively short-
duration activity. Female laryngeal fibers are primarily *slow-twitch*
or Type I muscle fibers and have greater endurance but contract
more slowly. The difference between the two is based on developmental exposure to sexual hormones. Exposure to greater
concentrations of androgens (of which the best-known is testosterone) during development changes the type of muscle fiber
that will be expressed. However, sexual differentiation of the
larynx is not the driving force behind differences between male
and female frog calling—it is more of a co-effect. What drives

the differences between male and female calls are the sexually differentiated vocalization regions of the brain.

Sexual differentiation of the brain has been a contentious issue in science for centuries. Early nineteenth-century anatomists cited the smaller size of the average human female brain to claim that women were inherently less intelligent; later, scientists graduated to debates on whether gender orientation is genetic or behavioral. But across all these years, the connection between sexual behavior and the brain has remained about as complicated a topic as you get in science.* And so *Xenopus laevis*, with its relatively limited but sexually differentiated vocal repertoire, is a wonderful animal for studying the basics of an extremely complicated subject. For example, if a male *X. laevis* is castrated, it stops calling altogether after about a month, but if provided with testosterone, it will begin giving out sexually appropriate calls again. If adult females are given testosterone, they will begin giving faster and faster trill calls, not able to match the highest-speed calls of normal adult males due to limits on their laryngeal muscles, but definitely masculinized. This is a different and arguably simpler system than that observed in birds and mammals, which usually only change their calling behavior in specific critical periods.

To try to get some kind of objective handle on the basic properties of the sexually differentiated brain, Ayako Yamaguchi and colleagues figured out how to remove the entire brain of male and female *Xenopus laevis* and keep them alive and active for quite a while. With the removal of all the superfluous

---

* For a great overview of how weird the intersection of science and sex is, read *Bonk* by Mary Roach. You may need mind bleach after reading it, though.

external gunk that animals spend so much time worrying about, an isolated brain in a dish, kept alive by an oxygenated bath of artificial cerebrospinal fluid, has time to focus on the really important things, like sex. Yamaguchi and her colleagues found that if she applied serotonin, a neurotransmitter involved in many complex behavioral tasks, the brain would begin to produce what is called a "fictive song"—sending neuronal signals from a song generator in the brain down the laryngeal nerve at rates appropriate to the sex of the frog whose brain was currently on vacation from its body. Subtle sexual differences between distribution and function of serotonin receptors, specifically those called 5-HT2sc receptors, change the rate at which signals are sent down the nerve, yielding different songs from male and female brains.

But so far we've still been talking about underwater life, which, as your experience with listening to the radio while you're in the bathtub shows, is not the same as life that hears and communicates through the air. Here we move from the aquatic *Xenopus laevis* to ranid frogs such as the American bullfrog (*Rana catesbeiana*), a species that listens and makes noise both in *and* out of the water. Bullfrogs are the largest North American frog—I've worked with older females who have weighed over 2 pounds and had no trouble fighting me off when I tried to pick them up.*

Like *Xenopus laevis*, the American bullfrog's social behavior is centered on calling and being heard. But unlike *X. laevis*, these are animals that spend a lot of time out of the water (or at least with their heads out of water). A typical night in a bullfrog

---

* I always eventually won, but usually only after being showered with at least a pint of bullfrog pee. To which I am allergic. Score: bullfrog, 1; scientist without Benadryl, 0.

chorus, where hundreds of males may be sitting around the edges of a pond making their "jug-o-rum" advertising calls, can reach deafening levels of 100 dB or more, a seemingly chaotic wall of sound. But it makes sense to the frogs. These calls both advertise the males' territorial claims, warning off other males, and try to lure females to their clammy green embrace.

If you go out to a bullfrog pond on a summer night armed with a flashlight, you can get an instant cue to the differences hearing in air entails as well as a quick way to tell the difference between male and females—look at their ears.* Just behind their eyes, there are plate-like structures called tympani, the bullfrog equivalent of your eardrum. The female's eardrum gets to about the same size as her eye, half an inch or so. But in a male bullfrog, the eardrum keeps growing throughout the frog's life. A full-sized adult male's eardrums can be an inch and a half across, sometimes giving it the appearance of wearing rather flat studio headphones. This is not a sexual signal like a peacock's tail. The size of the tympanum gives you an idea of what frequency the ear will be most sensitive to—the larger the area, the lower the most sensitive frequencies.

A male bullfrog's advertisement call is bimodal, meaning that it has two frequency peaks: one in the lower region around 200 Hz, and one higher at around 1,500 Hz. This high-frequency peak is species-specific: it is relatively unique to bullfrogs, at least compared to other species that share a bullfrog's normal calling environment. Female bullfrogs' tympani stop growing at a point that leaves the tympani most sensitive to this species-specific 1,500 Hz peak. But the male bullfrog's ears keep growing, making them more sensitive to lower and lower frequencies as they

---

* Which will no doubt provide you with a great thing to talk about at parties.

age. The difference between the sexes is important. Female bullfrogs want to be particularly sensitive to the pitch of advertisement calls of other bullfrogs (and only males call). This lets them know they are actually approaching a male of their own species, and also lets them select males who would be fit mates. Any kind of vocalizing is energetically expensive, so a male frog that calls loudly and a lot is probably healthier than one who just gives out the occasional call and is probably a better choice for a mate.

But the males' tympani keep growing over the course of their lives, and in hearing, size does matter. A larger receiving surface such as an eardrum is more sensitive to lower-pitched sounds. This will let the bullfrogs detect other calling males at greater distances, since higher frequencies attenuate faster than lower frequencies, and can let them determine the distance to potential competitors.

An advertising call is much more complicated than a simple "hey baby" let loose in a bar. Despite the fact that their brains are small, males still have to cope with a complicated acoustic environment, so there are rules. While the quiet females either swim between the calling males on the water's edge or hop overland from a pond with less action, a male hearing another male across a pond will wait until that one has finished its string of individual croaks made into long calls and then answer. Males who are close by will either shut up and stay out of it or, if they think the caller sounds smaller or quieter than them, will try to outdo them. This often leads to aggressive calls from the nearby neighbor, and if no one backs down, it can result in the deadly seriousness of two males battling over space, little green sumo wrestlers intent on defending their piece of swamp, even if it requires that they swallow their opponent's head (which is as big as their own). Meanwhile, with sneakiness that would do Machiavelli proud, smaller males who hang around the big callers, called satellite males, will intercept any approaching females to mate with them while the big boys are proving their points to each other. Even in the frog world, sometimes brains beat brawn.

When I first started studying frogs, one of the big questions we were working on was how frogs perceive pitch. Pitch is the psychophysical correlate of a sound's frequency: higher frequencies are heard as higher pitches. It's based on the brain's interpretation of the neural representation of the sound. Frogs can hear only relatively low frequencies—only up to about 4 kHz, compared to humans' 20 kHz. Their auditory nerves send signals to the higher centers in the brain using *temporal coding*, in which the firing of the auditory nerve is synchronized with the timing or phase of the sound. A neuronal signal, called a spike, is usually about 2 milliseconds in duration, or 1/500 of a second. This is the neuronal equivalent of a computer's or MP3 player's *sampling rate*. This means that for low-frequency sounds, 500 Hz and below, auditory nerves from the inner ear up through the frog's midbrain can spike every time the frequency hits the same point in its cycle or phase. But since neurons are biochemical signalers rather than silicon chips, with rare exceptions they can't fire in synchrony with every spike for frequencies much above 500 Hz (about the fundamental, or lowest, frequency of a young child's voice), so groups of neurons fire at mathematically related intervals called volleys that allow the brain to code sounds up to about 4,000 Hz.

This sounds like it would handily explain bullfrog hearing as being a simple model for pitch perception—they make low-pitched sounds and have good hearing across a decent range. But the advertisement call of male bullfrogs demonstrates one of the issues that comes up with psychophysics: sometimes you hear something that is critically important to you, but the physics shows it isn't there. If you look at the spectrogram of a bullfrog call, what you see is a series of lines in different frequency bands, starting at about 200 Hz and petering out somewhere at about 2,500 Hz. The lines are spaced almost regularly but with gaps between them in a pattern called *pseudoperiodic.* Most of the lower-frequency bands are spaced about 100 Hz apart: 200 Hz, 300 Hz, 400 Hz. If you were to take a simple mathematical approach to what this call would sound like, you would say it would be a low-pitched tone with the bottom end at about 200 Hz (which is about the fundamental frequency of an adult human female voice), but rich in harmonics, giving it a complex fine structure or timbre. But what the frog actually *hears* (based on recordings from frog brains and auditory nerves over the decades) is a tone that is primarily 100 Hz. How can you get a 100 Hz tone when there is no acoustic energy at that frequency? Because even the small frog brain uses a principle called the *missing fundamental.* If the harmonics of a call are spaced at regular intervals, the brain correlates the differences between the spectral energy bands, calculates the timing or period of this difference, and "hears" what isn't there—a pitch of 100 Hz.

This may seem a bizarre oddity particular to frogs, but it works just as well in humans, and it's the basis for some of our most common sound technology, including telephones and low-cost speakers. Most telephones have very small speakers, unable to reproduce frequencies below 300–400 Hz, yet it's pretty easy to recognize an adult male voice, which typically has a fundamental frequency between 150 and 200 Hz. In addition, most inexpensive audio or computer speakers, even with small subwoofers, don't perform well below 100 Hz. The reason you are able to "hear" the low pitch of a male voice on a phone or a good deep bass line on modestly priced speakers is that your brain is using the neural computation bestowed on us by evolution to fill in the gaps in the hardware's abilities.

But the aspect of frog hearing that fascinated me from the start was how they "learn" to hear. Frogs, like all amphibians, lay their eggs in water. Their young, called tadpoles, are radically different from their parents. Bullfrogs are partly terrestrial four-legged carnivores, virtual lions of their scale and their environment, with quite noticeable external ears. Adult bullfrogs have two auditory pathways to get sound to from their ears to their brain. One is a vibratory pathway that picks up very low-frequency sounds from the sides of their body, from their head, and from the ground through their forelegs and passes them to their shoulder girdle, to a muscle that connects to a piece of cartilage over the oval window, which leads to the inner ear. This opercularis pathway (named for the opercularis muscle, connecting the shoulder to the inner ear) is analogous to human bone conduction—it relies on passing vibrations through rigid elements of the body rather than dedicated external hearing structures. The second sound path, called the tympanic pathway, is the one more similar to what's found in our ears, attaching the external eardrum to a bone-like structure called the stapes, which connects to the front part of the oval window of the inner ear.

But tadpoles are limbless, totally aquatic herbivores with no visible signs of ears, and this has caused a lot of problems for scientists. Bullfrog tadpoles can spend two years developing before they become froglets, then live seven years as adult frogs. When the tadpoles hatch, their inner ears look very fish-like— they have large, prominent saccules, and two smaller otolith organs: the utricle, which is sensitive to lateral motion and is part of the vestibular system, and the lagena, which responds only to very low-frequency, mostly vertical vibrations. The pressure-sensitive amphibian and basilar papilla, the evolutionary analogues of our cochlea, only develop later on, as the tadpoles progress toward froghood. Even at this stage of development, the lack of external ears seems odd for a species so dependent on hearing for breeding, and anatomical studies of tadpoles showed no sign of the auditory periphery seen in adults. And while tadpoles have lungs even at hatching, there was no sign of the kind of swim-bladder-based specializations that give some fish good hearing. This led many to suspect that tadpoles were deaf or close to it. The only connections to their inner ears are through a bizarre strand of connective tissue called the *bronchial columella*, which connects the back of the inner ear to the lungs and actually passes through the aorta. Connecting the lungs to the inner ear would seem to indicate that if tadpoles hear, their hearing would be rather poor. First, this bronchial columella is not formed of bone or cartilage—it is composed of fibroblasts surrounded by collagen and has all the structural strength of al dente linguini. Therefore, any vibrations passed along it would likely be distorted. In addition, if tadpoles did get their sound via this wobbly structure, not only would they have to put up with a constant pulsatile sound from their own heartbeats, but they would have variable hearing based on whether their lungs had air in them or not.

The theory that tadpoles were basically deaf held for about forty years before anyone even tested it. The first attempt to record auditory responses from a tadpole's brains showed that the tadpoles had the expected poor hearing sensitivity. Case closed, it would seem. However, rather than establishing a scientific fact, this study highlighted one of the problems scientists face when studying things based on expectations rather than on testing basic facts: the tadpoles they recorded from were wrapped in wet gauze on a board out of the water and had airborne sounds played to them. Imagine that a frog scientist was trying to test your hearing while your head was underwater in a bathtub. The results would indicate that you have very poor hearing, with almost no responses to low-frequency sounds (as shallow water acts like a filter for higher-frequency ones) and a complete inability to localize where sounds were coming from. To the frog scientist, you are clearly deaf.

When you want to find something out about an animal's behavior, it is critically important to test it in a setting similar to its natural environment. Admittedly, this is very difficult—it is hard enough to carry out electrophysiology with the animal in a normal soundproof booth, and trying to keep an electrical system running with the degree of delicacy needed to record individual neural responses while keeping the animal's head under water is almost impossible. So, of course, decades after the issue was pronounced solved, I had to try it.

I went through massive amounts of aluminum foil (for grounding a pool of water), duct tape, and Tupperware containers to make a customized underwater recording tank, and it took me quite some time to figure out how to expose the tadpole's brain but not let the water into the opening (as well as how to be delicate enough with the surgery to make sure the tadpole could wake up and continue its development towards froghood).* But when I'd done all that, I found out that about sixty years of supposition about tadpoles was wrong.

---

* I also had to make customized tadpole earmuffs to block the sound but not squeeze their squishy heads. Patent pending.

Tadpoles in fact have excellent underwater hearing. But even though they live underwater for most of their development, they are not fish and could not be tested the way you'd test fish. Early-stage tadpoles hear much in the way sharks or simple fish do, with the sound passing through the tissue on the side of their head and impinging directly on the oval window to transmit vibrations to the saccule and other developing organs in the inner ear. Later-stage tadpoles, who have both hindlimbs and forelimbs, have a functional low-frequency opercularis pathway from their sides and forelimbs to their inner ear, although the tympanic pathway doesn't appear until about twenty-four hours after they absorb the last of their tails to become froglets. The problem is that when I was trying to record from some tadpoles, I was getting nothing. Zip.

After about ten of these trials, I was pretty sure I was not getting faulty results, so I went to my advisor. At first she gave me The Look. The Look can melt hostile deans at 40 meters, and I was just a grad student. But when I started going over all the data with her, we both noticed something odd: all of these "deaf" tadpoles were from one very short period of development, just before their front legs emerged. It turns out that in this brief period, about forty-eight hours long, while the low-frequency pathway is developing, the pieces of cartilage and muscle that attach the inner ear to the shoulder girdle block the opening on the side of the inner ear, the oval window, that let sound in when they were younger. In getting ready to move to a life where they have to hear vibrations from the ground, and eventually sounds in air, they undergo a brief "deaf period." At the end of that forty-eight hours, their hearing suddenly returns, with a broader range of frequencies and better hearing at the low end. Over the next week, they continue developing into froglets, and their tympani emerge on the side of their heads as they continue their journey to hearing in air; at this point their hearing is more sensitive to higher frequencies than in adult frogs, but they are ready for a truly amphibious auditory life.

Aside from the fun of overturning more than sixty years of scientific misconceptions about tadpole hearing, why do research like this? I'm sure a lot of people ask, "Who really cares about tadpole hearing? What does it have to do with me?" Quite a bit, actually. We humans too begin life as underwater organisms. We spend the first nine months of our life floating in the shallow pond of our mother's uterus, largely ignorant of anything but our most immediate surroundings. But at the beginning of the third trimester, our developing brains connect up with our developing ears and we begin hearing. As in any shallow pond, it's noisy but oddly muted, immersed in the constant rhythms of our mother's heartbeat and breathing, the lower frequencies of her voice filtered through the shallow uterine waters. As with the fish in an aquarium (or your adult self underwater in the bathtub), few sounds penetrate the air-water interface of the abdominal and uterine walls, and those that make it through are muffled, the higher frequencies filtered out by flesh and fluid. Sound provides the fetus its first taste of life in the greater world.* But upon birth, the muffled world of the womb gives way to a cacophony of sounds previously unheard. And like the developing tadpole getting past the deaf period, we too suddenly have to use our new ears to hear sounds in air, rather than picking up

---

* For a charming example, see Bruno Bozzetto's animation *Baby Story*—particularly at about 7:56, when the mother goes dancing.

low-frequency sounds through our skulls and fluid-surrounded ears. No wonder one of the first things we do after taking our first breath is cry. It's noisy out here.

But aside from being an interesting and useful model to study the transition from fetal to newborn hearing in humans, the story of tadpole hearing provides a hidden bonanza. Before a tadpole develops forelimbs, its brain has a rather fish-like connection pattern. It has an auditory nerve that carries signals from both the hearing and balance parts of its inner ear, and it has a lateral line system that sends signals through a separate pair of nerves projecting into a region of the hindbrain called the dorsal medulla. Here the nerves segregate into separate regions, some for processing balance and vibration, some for processing sound, quite a few that combine the two, and some for lateral line. All of these cross-connect in a regular pattern, which allows for comparison of signals from each side of the tadpole's body. This helps it do things like figure out where sound is coming from (using a brain nucleus called the *superior olive*), maintain its position in the water, or even just startle and swim away from danger (which can include its own parents—adult bullfrogs are rather indiscriminate in their choice of prey). Many of these brain regions then feed forward to the auditory midbrain, called the *torus semicircularis*, where auditory signals are recoded to handle complex sounds, and inputs from multiple sensory systems, including vision, are integrated to be sent forward to what passes for decision-making regions in the tadpole.

But when the tadpoles enter the deaf period, something changes. We still don't know why, but at the same time that the new auditory pathway blocks the inner ear from receiving underwater sounds, the brain rapidly rewires itself. The medulla disconnects from the midbrain, the superior olive drops out of the circuit, and the lateral line system begins dying entirely. There is a sudden upsurge in growth proteins, and brain regions begin moving around, reconnecting, and undergoing massive chemical shifts. About forty-eight hours later, as the new auditory pathway becomes complete and starts letting sound in again, the brain reconnects itself with a remarkably different configuration, one more suited to hearing in air. Within forty-eight hours, the tadpole basically rewires about a quarter of its brain.

This discovery was made in 1997, and since then some of my colleagues and I have worked on more than a dozen projects related to trying to understand how this transformation works, using everything from gene screening and sequencing to just watching a tadpole's behavior as it transforms into a frog. We've confirmed that frogs are critically important to our understanding of hearing and the brain itself because of their incredible ability to reshape their brains—not just in the course of normal development but even after injury. A frog doesn't just shift its brain around according to its developmental program; a frog's brain *heals*.

Each year millions of people are deafened, paralyzed, blinded, or rendered mute, not because of damage to their limbs or ears or eyes or mouths but rather due to irreversible damage to their brain and spinal cord. While humans can regenerate peripheral nerves, central nervous system damage is usually permanent. Frogs, on the other hand, are able to overcome it. Studies by Harold Zakon and others demonstrated that, unlike humans, if a bullfrog suffers damage to its auditory nerve, which would lead to permanent deafness in humans, the bullfrog not only heals but re-forms appropriate connections to allow restoration of function. Loss of sensory hair cells in the inner ear, whether due to injury or to exposure to certain types of antibiotics such as gentamicin, is typically permanent in humans, but not only do frogs regenerate after injury and drug exposure, there is evidence that they continually create some hair cells as old ones wear out. Somewhere along the evolutionary chain between amphibians and mammals, we lost the ability to heal our brains, our cranial nerves and much of our ears. So studying how a tadpole can completely rewire its brain in forty-eight hours and how a bullfrog can regrow its auditory nerve and restore function is not just basic research done for the sheer fun of doing science. These studies are likely to provide the clues that may allow us to create gene or pharmacological therapies that restore this ability to humans.

Chapter 4

THE HIGH-FREQUENCY CLUB

W HEN I WAS three years old, I went deaf. No maternal exposure to rubella, no overly vigorous toddleresque Q-tip exploration, just an unfortunate case of chicken pox that lesioned my eardrums. I don't have any explicit memories of the incident, and my hearing returned, the only residual problem being that my eardrums were slightly scarred and thickened. But now I hear bats.

Most people avoid bats if they can. Even on those warm summer evenings when you are outside, most of your interaction with bats is probably limited to seeing their shadowed forms flittering about, the smaller ones saving you from mosquito bites and the larger ones saving your garden from junebugs. But I spent a great deal of time one-on-one with them in the lab, and it never failed to amaze me how an animal whose brain is the size of a peanut actually builds most of its world with sound, creating three-dimensional images from subtle shifts in echoes.

To most people, bats' auditory world is so far outside of the human hearing experience that they seem like silent shadows.

But bats and humans share a lot of genetic heritage just on the basis of our being mammals. And as a mammal, you're a member of an exclusive evolutionary club: the high-frequency club. If you've ever wished you could hear a dog whistle, console yourself with the knowledge that humans, like all other mammals, have a remarkably wide range of hearing. Non-mammalian vertebrate hearing is generally limited to an upper end of about 4–5 kHz (although some specialist birds such as owls and cave swiftlets can hear up to about 12–15 kHz). Of all the other vertebrates depending on sound to let them hunt, mate, define their territory, or avoid predators, we have the broadest hearing range, from the infrasonics of elephants to the 150 kHz natural ultrasound used by dolphins.

This isn't because we're more advanced or the new kids on the evolutionary block. Mammals are as old as the dinosaurs, but our ancestors at the time were rather mouse- or shrew-like, and likely to be listening sharply to avoid becoming someone else's food. It's just that our ears are more specialized than those of fish, frogs, reptiles or birds. And while ours shares basic features with all other vertebrate auditory systems, ours has two features that none of the others has, and both of which are critical to high frequency hearing—an outer ear and a cochlea.

Even among auditory scientists, there is a tendency to take the outer ear for granted.* For humans it's something to hang a pair of glasses on or to make fun of if they stick out too far. And since an awful lot of our listening these days uses earbuds or earphones that sit inside or cover our external ear to let us hear more clearly in noisy environments such as airplanes or the gym, we

---

* The outer ear is also called the pinna, concha, or auricle, depending on your background, age, and degree of pretentiousness.

tend to ignore the outer ear as a vestigial organ that doesn't do much (outside of featuring in a memorable scene in the movie *Reservoir Dogs*). But our outer ear is actually a fascinating evolutionary development that tells us a great deal not only about the environment in which mammals listen but also about what we listen to. And if you take your earbuds out and move your head around to listen, you can actually get an idea what the outer ear does.

The outer ear is basically a flattened cone of relatively stiff flesh ending at the entrance to the ear canal or *external auditory meatus*, the place past which the Q-tip shall not go. If you look at most sources that talk about its function, the pinna is described as a sound-gathering device for low-frequency sounds, increasing the volume of vibration-bearing space that can funnel sound into the ear canal and increasing the relative gain of sound by up to 20 dB. This alone would make it an impressive passive listening aid; remember that dB are logarithmic in nature and every 6 dB is a doubling of the sound pressure, so a 20 dB gain gives you a lot more sound to work with. (If you look at antique hearing aids from the nineteenth century—or old cartoons—you usually see what was called an ear trumpet, a device that looks like a long metal horn with the small end fitted into the ear. It was basically an enlarged prosthetic outer ear.) But this change in gain only applies to sounds below about 4–5 kHz, the range at which all other vertebrates hear quite well.

High-frequency sounds, on the other hand, get horribly attenuated in the air due in part to their short wavelength. They tend to get degraded into thermal noise over relatively short distances unless they are extremely loud. So to use them, you need something that gathers more sound than either a small hole in the side of the head or even an eardrum stuck out on the edge of your skull, as with frogs. You need the equivalent of an audio telescope lens, something with a larger sound-gathering area. And not only do you need something to gather more sound, but you have to have some way of discriminating subtle changes in the sound based on the direction the sound is coming from, hopefully without having to swivel your head back and forth constantly to try to figure our where that high-pitched sound is coming from.*

Think about what the outer ear actually looks like—look across the room at someone else's ear or gently feel your own. It's not just some simple cartoonish cone. It is an extremely individualized shape, replete with ridges and usually a small tab that points slightly forward just over the opening to the ear canal (this structure, the *tragus*, is usually much larger in smaller mammals). For higher-frequency sounds, particularly those above about 6–8 kHz for humans, the ridges and valleys act as little blockades that slightly reduce the amplitude of certain frequencies of sounds, creating one or more spectral notches. The pattern of notches in the spectrum of the sound is specific to the shape and position of your ear. This "pinna notch" helps you localize high-frequency sounds, especially in the vertical plane. A sound coming from above or below your head hits these ridges and the tragus at different angles, with the result that different frequencies are slightly blocked. You can check this out yourself, especially if you go outside on a summer day when there are cicadas or other loud insects around. If you hold your outer ears flat against the side of your head (including the little flap in front of the ear canal (without, of course, blocking the canal itself), you'll find it's

---

* Although some mammals tend to do this as well.

much harder to figure out if a sound is coming from above, be-low, or level with your head.

The pinna notch has another interesting function, one that to my knowledge has not been ever studied, though it's rather obvious once you spend too many years thinking about things auditory. It just requires watching a mammal listen to a new sound that may be of interest to it—dogs are great ones to try this out on. Find your subject, be it a dog, a small child, a kitten, or a roomful of students, and say some nonsense words at it, but intone them as if you are saying something meaningful, like "Do you want a treat?" or "This will be on the final." What do they do? They roll their heads slightly to one side or the other (humans tend to roll their heads to the left, I've found). I've never actually seen this written up in any experiments, but in the course of torturing friends, family, students, and pets in the name of auditory science, I've found that it's pretty consistent—so much so that if you look at comics or cartoons of a mammal trying to figure out something it's just heard, you'll often see its head tilted to one side. By tilting the head this way, the listener shifts the position of the outer ear and changes both the timing and spectral properties of the sound, which allows the listener to hear it slightly differently when it gets repeated. It's sort of an auditory equivalent of 3-D movie glasses—rather than seeing slightly visually shifted visual scenes, by tilting your head you hear the sound from a slightly different auditory position, which both gives you more information about where it's coming from and lets your brain confirm what you heard.

Given that mammals have this specialization that would let them gather and tweak high-frequency sound, how do we act on it? The sensitivity of the middle and inner ear of other vertebrates largely tops out at about 4–5 kHz, but we have an evolutionary adaptation of our inner ear, the cochlea, a snail-shell-shaped structure full of sensory hair cells connected to the outer world via the auditory bones of the middle ear and the eardrum. That description is the same general plan of every other vertebrate inner ear as well, but the cochlea is significantly more complex. In the cochlea, the tips of the hair cells are embedded in a *tectorial* or "ceiling-like" membrane, and the base of the cells is in a long, thin trapezoid-shaped membrane called the basilar membrane. The shape is important—the basal end near the oval window, closest to the outside world, is narrower and stiff, and hence vibrates the most in response to high-frequency sounds. The far or apical end is wider and looser, and more responsive to low-frequency sounds. This variable flexibility causes the hair cells in different regions to vibrate maximally in response to a particular frequency range.

Because of this arrangement, hair cells don't have to fire in precise synchrony with the timing of the sound's phase. Instead, the hair cells' tuning is defined by their placement on the basilar membrane. Sound enters the fluid-filled chamber of the cochlea and creates a *traveling wave* with maximum deflection at the place on the basilar membrane corresponding to the particular frequency. This lets the sensory hair cells respond by *place coding*—it relieves the auditory nerve from the burden of trying to fire tens of thousands of times per second.

The traveling wave in the cochlea was discovered not by studying animals but rather by studying human cadavers, and won Georg von Békésy the Nobel Prize in Physiology or Medicine in 1961. The problem is, his theory turned out to be at least partially wrong, as it couldn't explain how complex sounds would actually break up into their component frequencies as they traveled through the cochlea. This illustrates a basic problem in anatomical science: dead preserved tissue doesn't work the same way as living tissue, especially in the case of something as dynamic as hearing. Dead guys not only tell no tales, but they also don't hear so well.

The mistake in von Békésy's theory was revealed by looking at another mammalian specialization of the inner ear that is not identifiable in dead tissue. Mammals have an additional set of cochlear hair cells called outer hair cells. For every inner sensory hair cell, there are three outer hair cells. And while sensory hair cells bend and deflect in response to vibration of a specific frequency, outer hair cells do something different. Like the inner sensory hair cells, the tips of the outer hair cells are embedded in the tectorial membrane, while their bases are in the basilar membrane. However, outer hair cells have tiny molecular nanomotors, similar to the mechanism in muscle fibers, that can pump up and down. When a sound comes in and vibrates the underlying basilar membrane, the outer hair cells amplify the signal by pulling and pushing on the upper membrane in synchrony with the sound vibration. This action also damps the vibrations in parts of the basilar membrane that have lower energy vibrations. So mammals not only evolved a physical structure that gave us more acoustic range but also evolved a series of frequency-specific internal amplifiers, the two interacting to give us the greatest hearing range of any other vertebrates.

To paraphrase Stan Lee, with great range comes great diversity. For example, I was doing a study in which I wanted to compare hearing in bats to hearing in mice. I was working with a very specific type of bat (*Eptesicus fuscus*, the big brown bat that is likely the most common bat where you live). Big brown bats are loved in auditory science because they actually form three-dimensional images of their world with echoes using biosonar. Yet at another level, their hearing is not too different from that of other small mammals. If you compare the audiogram (a measure of sensitivity to different frequencies) between a common mouse (*Mus musculus*) and a big brown bat, they hear just about the same frequency range. The difference lies in what they do with it. Mice are little furry bundles of fear—they are basically on everybody's menu, from praying mantises to your cat—so their hearing is very sensitive to a wide range of frequencies, allowing them to detect the slightest noise and run from it. Bats are much harder to catch and so use the same frequency range to detect changes in the echo from their 100+ dB calls with such sensitivity that they can tell the difference between a junebug and the leaf right behind it while flying at 25 mph in total darkness.

The problem comparing the two arose when I did something I do every day and typed "bat," "mouse," and "audiogram" into PubMed, the search engine for the biological sciences. The question became, which mouse? There are literally *hundreds* of different strains of mice available, many with interesting custom mutations (such as going into seizures when exposed to the wrong sound), many with side effects of these mutations (such as going deaf after four to five months). While this has created a bonanza for modeling different aspects of hearing (and many other factors), it starts begging the question of what a mouse hears. You have to ask not only what mice hear but also which mouse. More interesting to me, if a mouse is considered a generic mammalian model, how did some mouse-sized shrew-like creature 40 million years ago start an acoustic arms race to become the auditory Terminator that is an echolocating bat?

Bats have taken hearing to its most extreme form. Bats'

nighttime hunts are high-performance aerial dogfights carried out in total darkness, replete with high-speed chases, moving targets, and biosonar tracking systems—only the bats eat their targets. So while bats may not be the closest animal models to study human hearing, they are the best model for the other type of translational research, such as the development of bio-mimetics, which produces cutting-edge technology based on natural forms. Starting in 1912, when Sir Hiram Maxim drew on bat behavior to propose outfitting ships with active sonar to act as a "natural sixth sense," military organizations around the world have poured millions of dollars of research into under-standing how bats normally do things we wish a human fighter jock or an autonomous drone could do.* So how do you study sound production and hearing in an animal whose lowest-pitched calls are barely audible and go up to frequencies five times higher than the upper range of human hearing?

There's a room in the basement of the Cognitive, Linguistic and Psychological Sciences Department at Brown University that is unnerving to most people. About 40 feet long, 12 feet high, and 15 feet wide, its walls and ceiling are covered with black acoustic foam and the floor is heavily carpeted. It has an inde-pendent air supply with virus filters, with a control switch on the wall to shut down the blower. Within this room is another room built of copper mesh, effectively electromagnetically iso-lating it from the rest of the universe. Every meter or so along the walls sits a tiny ultrasonic microphone array, feeding back to a twenty-four-channel audio mixer that ties into a computer

---

* Not just the military is interested—NASA has funded some of my bat re-search in the hopes of developing an autonomous flying robot explorer that could make its way through the smoggy skies of Titan.

with high-speed broadband data sampling cards capable of sampling sounds up to 100 kHz. Oh, and it's normally completely dark. The only illumination comes from infrared emitters that light the scene for specialized equipment that can later reconstruct the whole thing in glorious detail. The interior is filled with nets, rope, floor-to-ceiling plastic chains, little Styrofoam balls on sticks, ultrasonic microphones, and IR-sensitive video cameras, plus little chunks of mealworms and the occasional Necco wafer on the floor.* Once you enter and close the door and the lights, the silence and the dark are overwhelming. The acoustic foam sucks up all sound, so after a few minutes the loudest thing you hear is the banging around of air molecules in your ear canal until you eventually start hearing your own heartbeat and nothing else. And did I mention it's dark? This is not a scene from a steampunk Edgar Allan Poe play—this is the Brown University bat playground, under the direction of James Simmons.

The bats flown in this room are typically *Eptesicus fuscus*, the big brown bat. Bats have long been a creature of myths, most of them rather dark in nature. This leaves most people with the idea that bats are rare nocturnal creatures living on the outskirts of what we think of as a normal, daylit environment. But bats are the second most common type of mammal, with more than 1,100 different species in every continent of the world except Antarctica. They fill every non-aquatic terrestrial niche, from Australia's gray-headed flying foxes (*Pteropus poliocephalus*), a

---

* Bats don't like Necco wafers, but since the candy gives off echoes about as strong as a dangling mealworm, bat scientists have been using them to try to confuse echolocating bats for about fifty years. I personally don't know anyone who actually *eats* them.

daytime non-echolocating bat that eats pollen, nectar, and fruit, to the Central American vampire bat (*Desmodus rotundus*), which not only is a nocturnal echolocating blood feeder but also has a pit organ like a viper, giving it thermal vision.

Nevertheless, most bats are nocturnal insect eaters. Some, like the pallid bat (*Antrouzoius pallidus*), feed by gleaning—listening with very elongated ribbed ears for the sounds of scorpions or other insects as they scuttle through the sands. But most of them rely on echolocation or biosonar, sending out signals whose echoes cue the bat in on the location, distance, and nature of the target. As bats typically eat their own body weight in insects and other arthropods every night, they need an incredibly efficient system to locate and grab their prey while simultaneously not flying into less edible things such as trees and researchers' nets. These nocturnal bats are *not* blind. Their vision is in fact comparable to that of any other small, twilight-active mammal, and some have suggested that certain migrating species can even navigate by the brighter stars. However, low-light mammalian vision is reasonably good for motion detection but notoriously bad for resolving form, especially when the viewer is pulling 9 G turns at 25 miles per hour.

Echolocating bats use sound differently than we do. We are passive listeners—we listen to sounds in our environment and try to identify what made the sound (based on its frequency) and how far away and where it is (by its loudness and phase). But bats are active listeners—they supply the sound they use to navigate their environment and listen for the echoes. Even though their brains are tiny compared to ours, they are mighty auditory engines. They compare the echoes to an internal representation of their own call and figure out what is out there based on tiny differences in those echoes stemming from what they bounced off of. Echolocating bats use two different basic types of calls, which generate different types of echoes. Constant-frequency (CF) bats send out calls that are mostly a single steady tone, sometimes with a small downward chirp at the very end. When a CF bat is flying around in a cluttered area, such as woods, the echo that returns is largely a delayed version of the tone the bat set out. But if the CF bat's tone strikes a fluttering insect, the motion of the insect's wings will create a Doppler shift of the echo, and the bat knows that something is moving in front of it. The other type of bat, including the big brown bat, is called a frequency-modulation (FM) bat. FM bats put out a different type of call, "chirps" that sweep from high to low frequency. FM bats call with more variability than CF bats: when they are just flying about, they put out chirps only about once a second, but once they detect an echo, they sharpen their calls, putting them out faster and faster and sweeping their echolocation signal around like a spotlight until they start receiving echoes. Rather than relying on Doppler shift to detect prey, FM bats work more like specialized radar units, picking up multiple reflections from targets and integrating them into the auditory equivalent of three-dimensional views of the world.

After several hundred years of experimentation, we have some idea of how bats do this.* At a basic level, they determine the distance to a target by translating echo delay into target range; this tells them how far away a bug or branch is, and they

---

* Lazzaro Spallanzani first postulated in the 1770s that bats navigated by sound, though he used techniques that would not make it through a review committee these days.

are amazingly accurate.* Parts of their auditory midbrain are best described as echo-ranging computers, comparing the time when neurons respond to a call to the time the bat emitted that call, then converting the time delay or latency into a representation of the distance. Part of this conversion is based on the actual time of delay, but as high-frequency sounds in particular tend to drop off at a nearly fixed rate, the bat's brain also assigns distance based on the loudness of the echo, a phenomenon called amplitude-latency trading. Bats can respond to changes in sounds of less than a microsecond, which is one-millionth of a second. This seems extremely counterintuitive, if not flat out impossible (which a few bat scientists still claim it is). All nervous systems run on the millisecond scale, seemingly fast enough for you or me. But experiments have shown that if you train a bat (and yes, they are very trainable) to respond to a specific echo delay— basically the equivalent of asking a person to read something a certain distance away—bats can tell differences in the delay of two sounds in the range of a few hundred *nanoseconds*. Nanoseconds are billionths of a second, so bats are detecting auditory features about a thousand times faster than your brain supposedly operates. This has led to years of serious arguments about the validity of the experiments, arguments that ended only after double-checking equipment calibration to cover elements such as the transit times of electrons in a cable of a specific length and absolute amplitude calibration at the subdecibel level, all this using equipment so precise that it requires manual programming in binary to set the time delays. According to classical neuroscience concepts of how the auditory system worked, the idea

---

* Humans are not, as my previously described classroom experiment showed.

seemed so farfetched that fistfights have broken out at scientific conferences over the interpretations of the findings. Bats would have to be some sort of superorganism to be able to have that level of precision. And yet they do it.

If it was just a question of getting a single echo back, it would be hard for a bat to distinguish between a tasty moth and a leaf, or worse yet, another bat. The structure of the bat's call, two harmonic bands sweeping downward from about 100 kHz to 20 kHz, provides the bats with a basis for getting echoes from objects of different sizes depending on the wavelength of the frequencies in the call, from about 0.3 to 1.7 centimeters across. In addition, because complex shapes such as an insect will reflect the call from different points on its body, the bat gets multiple individual echoes or *glints* with very slightly different echo delays. These glints change the fine structure of the echoes, and the bats are capable of using these to reconstruct the shape of objects, particularly when the object is moving and changing its relative shape. It's also why the bat speeds up its calling behavior and shortens its call to get more echoes and more information from the target as it gets closer, ending with a terminal buzz, hopefully right before catching its prey.★

Most studies of these basic parameters of bat echolocation have been carried out in the laboratory—for the most part, bats have been studied to further human technological causes. Sir Hiram Maxim's invention of sonar for submarines was based on

---

★ Bats do not catch bugs in their mouths. That would be like trying to trap something with your eyelids—it's going to block your field of view. The insect is usually snagged by a wing or their uropatagium (the tail and membrane around it) and thrown it into their mouth, with the bat often performing an aerial somersault in the process.

the idea (only vaguely understood at the turn of the century) that bats were using some kind of non-visual sensing, possibly using touch on their wing membranes to let them detect fine motions of air molecules. While the confirmation that bats were actually using ultrasonic emission only came about in the 1940s by Robert Galombos and Donald Griffen's invention of the ultrasonic microphone, known commonly as a bat detector, the idea of using an *active* sensory system drove the development of sonar and radar. But bats (at least the ones clever enough not to be caught) live in the real world, and the real-world auditory scene analyses carried out by bats are mathematical nightmares, especially for creatures whose brains are the size of peanuts.

Auditory scene analysis simply means having to deal with all the complex sounds that make up the real world, not the pristine acoustic environment of an auditory laboratory. All animals go through it, even frogs, who developed the aforementioned simple rules for when to call or when to pick a fight based on the loudness and pitch of their neighbors calls. But for bats it's much worse. If you're at a party and miss what someone says to you, you say "What?" and move closer. A bat, on the other hand, flying at high speed in darkness while chasing dinner, says "What?" and slams into a tree. Bats don't hunt too often in nice echo-proof rooms. They are flying in front of or within the branches of trees, each leaf and twig giving off an echo. And while they're somewhat territorial, bats often impinge on each other's territories, so they minimally have to know how to ignore another bat's hunting calls, or else have to evade an angry competitor bent on chasing it out of the rich hunting area it considers its own. Bats have to segregate echoes from their prey, other bugs out of range, and bushes and trees in their way, all the while ignoring calls from other bats. Experiments into how bats handle real-world

scenes have only begun over the last few years at the Riquimaroux and Simmons labs in Doshisha University in Kyoto, Japan,
and Brown University in Rhode Island, respectively. And it
turns out that bats are much more flexible than we thought.

As usual, the limitation for understanding the behavior of a
species with calls outside our own range was based on two factors. First is the problem of treating bats as sonar machines
rather than complex animals living in a complex world. This is
a normal if unfortunate outgrowth of translational research—
it's hard to keep all the actions of something as dynamic as a
mammal in mind when you're trying to represent them through
equations and hard data. The second problem is that to understand animals that hear outside of our range, we need to use
technological and experimental tools that change over time,
hopefully giving better results as they get more complex, rather
than just cluttering up the datascape. While the discovery that
bats could determine changes in echoes shorter than a microsecond happened in the early 1990s, and arguments about it raged
until just a few years ago, when better digital technology confirmed the findings multiple times, the underlying basis for bats'
temporal hyperacuity only began coming to light in the last few
years and required better molecular techniques than were available back in the day. Bats' ability to determine the timing of
echo structure turns out not to require superpowers—just a developmental feature probably caused by an evolutionary mutation that was retained longer in bats than other small mammals,
and only identifiable now, after great strides in brain research.

When the brain of a mammalian fetus is developing, it forms
connections based on patterns of gene expression. Our DNA
turns on and off numerous chemical signals that cause the cells
to differentiate into specific cell types following what's called

their *fate map* and migrate from the place of their birth to specific sites in the developing brain. But when the newly born and positioned neurons start connecting up, they lack the neurochemical complexity that makes the postnatal brain responsive to changing environmental conditions. Many developing neurons connect not with modifiable chemical synapses but rather with *gap junctions*, small channels that directly connect one neuron to another, allowing precise and rapid flow of signals between the neurons. This is useful for laying down early patterns of connectivity in the brain. But most of these gap junctions are replaced by chemical synapses as the brain develops, with only a few regions in the brain retaining them into adulthood. And none were ever seen in the central auditory system of a mammalian brain until 2008, when because of a shipping error, I received a small vial of antibody to connexin-36 (Cx36), the protein that makes up neuronal gap junctions, instead of the neural tracer I actually ordered. Antibodies are widely used in studying brain chemistry because they will latch onto a specific protein, and if you process them with a fluorescent marker, you can generate extremely colorful and precise maps of how that protein is distributed in the brain.

Rather than ship it back (it would have spoiled), I decided to try to see how Cx36 was distributed in a bat's brain compared to a mouse's brain. It turned out that for the most part, a bat's Cx36 distribution looked very much like a mouse's—the few regions that normally express the gap junction protein in adult mice were all lit up. But something looked wrong. The region where the auditory nerve first enters the brain, called the anteroventral cochlear nucleus (AVCN), was so brightly labeled that it was the brightest part of the brain, but only around the area that first contacted the entering auditory nerve fibers.

When I looked closer, it turned out that the label was in a very specific group of cells that seemed to form an interconnected network. This network was positioned right where they could affect the first auditory signals coming into the brain. Further experiments showed that they were cells not only full of connexins in their surface but also labeled with gamma amino butyric acid or GABA, the neurotransmitter that causes inhibition of other neurons. The clincher was that these cells, when labeled with a neuronal tracer, didn't actually project to the rest of the auditory system—they seemed to be set up to only act on either auditory nerve fibers or the first processing neurons that did send projections up through the brain.

We had found the biological equivalent of a temporal filter—something that lets only precisely timed signals enter the brain—and it's never been seen in other animals. The way we think it works is that the cochlea sends bundles of nerve fibers from regions that respond to similar frequencies. Since the mammalian ear is made of wetware rather than silicon and wires, there is a certain degree of slop in the time of arrival of signals from the cochlea, even from nerve fibers that may be only microns apart or infinitesimally shorter or longer than their neighbors. And while at the time of writing this book this is still under investigation, what appears to happen is that bats, by some evolutionary mutation that let them retain this fetal feature, develop a network-based filter at the lowest part of the auditory system that only allows the first signal from any given frequency to get through. This spectrotemporal gateway means that their auditory system isn't any faster than a mouse's or a human's—it's just much more precise. They don't have to be super-organisms or violate the law of physics to create images from sound; they are

merely the product of their particular evolutionarily driven development.

But what can we learn from bats that is translational—that
directly relates to human hearing? While bats have been the
inspiration for technological advances from sonar to ultrasound,
their ultra-precise hearing seems a form of evolutionary exotica
that has little to do with how humans hear. But it turns out that
bats may hold the secret to a problem that plagues all of us as we
age: presbycusis, or age-related hearing loss. We humans, like
all mammals, lose our hearing as we age, even if we didn't spend
years in front of high-powered speakers or under headphones or
working in loud environments. It's always been treated as a normal part of the aging process, but it underlies a lot of serious
cognitive and behavioral problems, such as the loss of ability to
understand speech in even normal acoustic environments, and
may be the basis for some of the paranoia associated with Alzheimer's and other dementias. It's easy to think that something
is going on behind your back when you can no longer monitor
the world out of your line of sight. The basic theory has been
that since the hair cells at the base of the cochlea (the narrowest
part, near the oval window) are closest to the outside world, they
take the greatest beating from acoustic input, whether "normal" sounds such as speech or potentially damaging noise such
as blasts or chronic exposure to subway noise. Since mammals,
unlike frogs, don't regenerate auditory hair cells under normal
circumstances, the high-frequency-sensing hair cells will be the
first to wear out.

But there is a serious problem with this theory. It's all well and
good to suppose that forty years of sound will start wearing out
the structure, but mice, who live only a year or two and are

one of the most common models for auditory function, start showing high-frequency hearing loss after about a year, with certain mutants showing symptoms in only a few months. So the issue is clearly more complicated than what a manufacturer would call normal wear and tear; there have been hundreds of studies examining various genes and gene products that seem to be involved (although not clearly causative) in high-frequency hearing loss. So we are looking at a complicated system that seems to be a universal problem in mammals. Or is it? The answer may lie in the bat's ear and brain.

Echolocating bats live an uncommonly long time. My favorite big brown bat, Melanie, was about sixteen when she died. There are documented cases of little brown bats (*Myotis lucifugus*) living to thirty-four years. This is ridiculously long for a small mammal with a high metabolism. A bat eats its weight in insects every day, and a hunting bat's heartbeat can reach 1,000 beats per minute—a rate that would make a human heart explode. If you follow normal metabolic models, bats should live three to four years. Even if you give them time off for hibernation, they still live three times longer than they should.* And yet these animals are absolutely dependent on high-frequency hearing. A deaf bat not only will slam into things but will starve to death because it can't hunt. Somehow, bats have evolved an auditory system that preserves at least the most critical range of their hearing necessary for echolocation, which for big brown bats starts at about 20 kHz and for little brown bats starts at about 40 kHz, and keeps it functional for years longer than any other known mammal could. And we have no idea why.

---

* In my next career, I'm going to figure this one out and live to 240 even if it means a steady diet of mealworms and moths.

Whenever I hear some student bewailing the fact that with the thousands of papers on hearing (and everything else in science) appearing annually, it seems like we've discovered everything, I just smile and shake my head and begin asking him or her all the questions we don't know the answer to. Do bats lose the extreme upper end of their hearing and just retain the low end, implying that this is merely a scaling function for their auditory range? Or do they have some molecular or systemic method for keeping their hair cells healthy? Or do they, like fish, frogs, and birds, have the ability to regenerate hair cells that are damaged? The list of things we don't know goes on for a very long time. Echolocating bats, which seem so exotic and strange, even to other auditory scientists and animal behaviorists, may hold the secret that will allow humans to have healthy hearing for their entire lives. We just have to figure out the right questions.

Chapter 5

WHAT LIES BELOW: TIME, ATTENTION, AND EMOTION

A FEW YEARS ago when I lived in southern Rhode Island, I went for a nighttime run. When I run during the day, I usually put on some headphones and run to something with a decent beat, partially to mask traffic noise and partially to give me an external basis for my pace.* But when I run at night, I leave the music off and just listen to the environment. The lack of sunlight makes the sounds much richer, and you're also safer being able to hear everything going on around you. I was on a downhill part near some favorite ponds, usually rich with the sounds of bullfrogs and green frogs trying to get lucky. But shortly after I turned onto the road and ran past the first of them, I noticed that they were quieter than usual, even after I had thunderfooted past. I kept running and started hearing what sounded like very soft footsteps, so I stopped, jogging in place, and looked around.

---

* Like James Gorman in his excellent essay "The Man with No Endorphins," I don't really like running so much as I like having run. It's sort of like beating your head against a wall—it just feels so good when you stop.

The road, aside from a single streetlight near a quarry a few hundred yards away, was dark, real *country* dark. Dark in a way that, as a recently transplanted New Yorker, I was just learning to cope with. But there was nothing there, and I didn't hear anything else, so I moved on toward the next pond. The frogs shut up on schedule as I got around a hundred feet away, but I thought I heard the padding sound again. At this point I was beginning to get a little nervous, so I did another stop and search; all was quiet. I decided that since I wasn't somewhere sensible like Brooklyn, where there are streetlights for illumination and bodegas to step into in the event of trouble, I would pick up the pace. Just as I was about to bolt toward the streetlight, I heard a loud splash and a snarl. I jumped several hundred feet straight up (okay, it might have been a few inches) and swiveled around to face the pond, my whole body tense and ready to run which-ever way the source of the noise wasn't. What I saw was the sor-riest, most bedraggled and shamefaced coyote imaginable. He must have been pacing me for the last quarter mile or so and, like his cartoon counterpart Wile E. Coyote, run right off the edge of the sidewalk and dropped straight into the pond, leaving him soaked in both water and the coyote equivalent of embar-rassment. He jumped out of the pond, shook himself dryish, and slunk off, muttering under his breath. I couldn't help myself—I said "meep meep" and ran off in the other direction.

This story shows why it's important not to wear headphones when you're in the dark: because hearing communicates more to you than just what's in your iTunes playlist, or even the sounds in your immediate vicinity. It lets you monitor the world around you even out of your line of sight and in the dark, and it does it faster than any of your other senses. Your brain is a pattern-seeking machine, constantly identifying relationships between all the sensations and perceptions that bombard you. Sensory inputs that are correlated in some way—by similar frequency, timing, timbre, or location (or, in the non-auditory world, shape, color, flavor, or smell)—cause neurons to fire in similar patterns at or nearly the same time. Neurons that fire in synchrony are more likely to trigger their target neurons to fire, hence passing the message "Something non-random happened" further up to the executive processing regions of your brain. Since your perceptions are based on binding common elements of sensory input together in time and space, other senses such as vision that are spatially limited and relatively slow often get false positives when correlating heavily overlapping or ambiguous features. This is why there are so many web pages devoted to cool optical illusions, and so few that even mention auditory illusions—it's harder to trick your ears. Hearing tends to be better at segregating inputs properly even though it gathers information from a much wider region, unlimited by line of sight. This is because hearing is faster than vision.

At first, thinking of hearing as faster than vision seems counterintuitive. We are used to assuming that our brain and vision are really fast—witness phrases such as "fast as thought" (about 750/1,000 second, according to one Johns Hopkins study) or "gone in the wink of an eye" (about 300/1,000 second). Sounds fast, and compared to the amount of time it's taken you to read this paragraph, it is. But it's all relative. Consider a few examples: the quartz crystal in an old-style watch oscillates at 32,000 times per second; our atomic clocks are set to the vibration rate of a single energized cesium-133 atom, more than 9 billion oscillations per second; iodine atoms vibrate 1 million billion times a second; and quarks exist for only a yoctosecond (10^{-24}

second), meaning you would have to go to 2×10^{-21} quark births and funerals in the time it took to notice that there was something stuck under your contact lens. Luckily, we are tuned to operate on a much more limited time reference scale, from a few tenths of a second to our own life span, which is, sad to say, only about 2.5×10^9 seconds even if you do go to the gym regularly.

But we are a visual species, diurnal (mostly awake during the daytime) and trichromatic (able to see a reasonably full range of colors), and we usually describe our surroundings or think about them using visual descriptions. Vision is based on the perception of light, and light defines the fastest possible speed in our neck of the universe, 300 million meters per second. Light from the surface of this page would travel to the back of your head where your visual cortex is in less than a nanosecond (one-billionth of a second). But our brains get in the way, and not just by blocking the back of our skulls.

Vision from input to recognition operates in the time span of several hundred to several thousand milliseconds, millions of times slower than the speed of light. Photons go in through our eyes and impact special photoreceptors in our retinas, and then things get slowed way down to chemical speeds, activating second messenger systems in the photoreceptors, whose signals pass through synapses to retinal ganglion cells. The ganglion cells then have to double-check with other retinal cells to see if they should respond to this input or ignore it, then accordingly fire or not fire a hybrid chemical-electrical signal down the long path of the optic nerve to one of several destination layers in the lateral geniculate. From there you have to synapse *again* and maybe go to the primary visual cortex or else the superior colliculus, but in either case, it becomes like the most hideous subway ride a poor little visual signal has ever been on.* And it takes hundreds of milliseconds just to get somewhere that might make you able to say, "Um, did I just see something?" Luckily for us, our brains are also temporally tuned to deal with "real time" based on vision. In real-world terms, changes occurring faster than about fifteen to twenty-five times a second can't be seen as discrete changes; thanks to numerous neural and psychological adaptations, we instead perceive them as a continuous change. This is very handy for the television, film, and computer industries, as they don't have to release drivers for a yoctosecond graphics card.

But hearing is an objectively faster processing system. While vision maxes out at fifteen to twenty-five events per second, hearing is based on events that occur thousands of times per second. The hair cells in your ears can lock to vibrations or specific points in the phase of a vibration at up to 5,000 times per second. At a perceptual level, you can easily hear changes in auditory events 200 times per second or more (substantially more if you're a bat). A recent study by Daniel Pressnitzer and colleagues demonstrated that some features of auditory perceptual organization occur in the cochlear nucleus, the first place in the brain to receive input from the auditory nerve, within a thousandth of a second of the sound reaching your ears. You figure out where sound is coming from in the superior olivary nucleus, which compares input from your two ears based on microsecond-to-millisecond differences in time of arrival of low-frequency

---

* As a matter of fact, some of the most complete visual connection mappings ever done, courtesy of Professor David Van Essen, look like the New York City subway map on steroids as drawn by Jackson Pollock.

sounds at the ears (or in subdecibel differences in loudness for higher-frequency sounds) and does this with only a few more milliseconds of delay. Even at the level of the cortex, which, as the top of the neural chain, is trafficking in enormous amounts of data from lower down and tends to be relatively slow, there are specialized ion channels that allow for high-speed firing and retain the most basic features of auditory input all the way up from your ears. So even after these signals go through about ten or more synapses to reach your cortex, where most of what we think of as conscious behavior takes place, it takes only about 50 milliseconds or less for you to identify a sound and point to where it's coming from.

But this is taking a classical view of the auditory pathway, such as you might find on Wikipedia or in an undergraduate class on the biology of hearing. At this level, it seems to be a pretty straightforward system. From the hair cells in the ear, sound is coded and passed along the auditory nerve to the cochlear nuclei, where the signals get segregated by frequency, phase, and amplitude, passed through the trapezoid body to the left and right superior olive to determine where the sound is coming from, and sent on to the nuclei of the lateral lemniscus, which processes complex timing processes. From there, the signals pass to the auditory midbrain, the inferior colliculus, which starts integrating some of the individual features into complex sounds, then sends them on to the auditory thalamus, the medial geniculate, which acts as the primary relay station to get things to the auditory cortex.

The auditory cortex is where you start becoming consciously aware of the sounds you heard all the way back in your ears, with specialized regions subserving different auditory specialties, such as the planum temporale, which identifies tones, and

Wernicke's and Broca's regions, which specialize in comprehension and generation of speech, respectively. It is also here that lateralization of sound processing emerges—in most right-handed people, the informational aspects of speech tend to be processed on the left side of the brain, whereas the emotional content of the speech tends to be processed on the right side. This lateralization leads to an interesting practical phenomenon. Think about when you talk on the phone. Do you put the phone to your right ear or your left? Most right-handed people put it to the right ear because the bulk of the auditory input from your ear crosses over to the left hemisphere to help you understand speech. This is sometimes reversed in left-handed people, but there are left-handed people with speech cognition on the left and emotional content cognition on the right. A similar variation can occur in right-handed people, too. I always found it weird that I have trouble understanding phone calls unless I listen with my left ear, even though I am right-handed. It wasn't until I volunteered in a neural imaging study years ago that I found out that I am one of the few mutants with flipped content/emotional processing centers. The rest of my brain is mostly normal, but there have been some studies showing that evolutionarily new areas have a greater tendency to show variation in things such as lateralization, and speech comprehension is one of the newest.

But auditory signals from the lower brain stem don't climb up to the cortex just so you can enjoy music or be irritated by the kid next door practicing the violin badly. Any single element of the auditory path has had dozens or sometimes hundreds of complex papers written about it, in a variety of species, ranging from connections of individual types of neurons within them to variability of expression of immediate early genes depending on the type of input you provide to them. As more and more data emerge about a system, you don't just add up more information; if you have the right mind-set (or enough grad students), you start realizing that the old rules about how things work aren't holding up so well anymore.

When I was a grad student in the 1990s, I still heard the terms "law of specific energy" (ears respond only to sound, eyes only to light, vestibular organs only to acceleration) and "labeled lines" (neural connections are specific to their modality—sound goes to the ears, which connect to auditory nuclei, and so sound ends up being processed as "hearing," whereas visual input goes in the eyes and becomes things you see). This concept of a "modular brain" was widely adopted—the idea was that specific regions processed the input they got and then communicated between themselves, with consciousness or the mind emerging as a meta-phenomenon of all the underlying complexity.

But starting around this time, neuroscience reached an interesting tipping point (to borrow a phrase from Malcolm Gladwell). Centuries of anatomical and physiological data were beginning to show strange overlaps. The superior colliculus, formerly thought of as a visual midbrain nucleus, brings maps from all the sensory systems into register with each other. The medial geniculate, while providing most of its input via a ventral pathway to the auditory cortex, also has a dorsal projection that goes to attentional, physiological, and emotional control regions. The auditory cortex can respond to familiar faces. The tens of thousands of published studies (combined with the availability of papers on the Internet, instead of deep in the science library stacks) started providing fodder for what has been called the "binding problem"—how all the disparate sensations get tied together via multisensory integration to form a coherent model of reality. Throw in the gene revolution in the last ten years, with sequencing, protein expression, and the availability of complete genetic libraries of organisms (at reasonable prices for those living on grants), and even more understanding of the brain's flexibility and complexity emerged, allowing insights into how inherited and environmental conditions can change how the brain responds. And once that particular genie was let out of the bottle, researchers started noticing that some of the things that formerly were addressed only through the old black-box psychological techniques could now be reexamined in a neuroscience context. This begins telling the tale of how sound not only lets us hear things but actually drives some of our most important subconscious and conscious processes. Which brings us to the paradoxical issue of why we don't usually pay attention to sound even though we use sound to drive our attention to important events in the world around us.

Let's go back to brain pathways for a minute. Auditory signals exiting from the medial geniculate do not necessarily wind up in parts of the brain scientists think of as auditory. They project to regions that traditionally have been lumped into what used to be called the *limbic system*, an outdated term that you still see in textbooks and even some current papers. The limbic system includes structures that form the deepest boundary of the cortex and control functions ranging from heart rate and blood pressure to cognitive ones including memory formation, attention, and emotion. The reason calling this a specific system is outdated is because the harder you look at the fine anatomical, biochemical, and processing structures of the regions, the less they seem to be discrete modules. They are more of an interconnected network with loops that feed both forward to provide information and backward to modify the next incoming signal. The problem of figuring out how sound affects all these systems based on anatomical projections is made more complicated by the fact that there are few direct auditory-only projections to these regions. To understand the complexities of the brain, sometimes you have to start looking at real-world behavior and work backward. So: what does sound have to do with attention?

A while back, I was contacted by the Perkins School for the Blind in Watertown, Massachusetts, to see if I could help with an acoustic problem. It seems that state-mandated fire alarms were panicking their students to such an extent that the kids would actually stay home on days when the alarm was to be tested. The staff wondered if it would be possible to make an alarm that would still get their attention but wouldn't terrify and disorient them. I started working on the problem (and still am), but the point of the issue was brought home to me recently when a fire alarm went off in the lab. After about thirty seconds of trying to get to my office through the halls, all of which were in plain sight, while simultaneously trying to turn my head to lessen the noise, I realized I was having trouble navigating. How much worse is it for people who not only can't see where they're going but have their cane taps and any verbal instructions masked as well?

An alarm—whether a loud bell, a klaxon, or a blaring synthesized voice yelling "Fire, fire" (like in my annoying kitchen smoke detector)—is a psychophysical tool. It presents a very sudden loud noise to get your attention (like my startle at the clumsy coyote's splash) and then continues repeating that signal. While you don't startle to subsequent loud sounds, the very loudness itself keeps your arousal levels high, and if the arousal isn't abated either by the siren ending or by you getting away from it, it's easy to have arousal change to fear and disorientation. This shows the linkages between sound, attention, and emotion.

At first it might seem that you couldn't pick more-different aspects of behavior. Attention gets you to focus on specific environmental (or internal) cues, whereas emotions are preconscious reactions to events. These two features seem remarkably different, but are both based on getting you to change your response to your environment before conscious thought takes over. The long, subtle buildup of arousal that occurs as you realize that something is wrong with your environment when you don't hear things you expect and the sudden onslaught of fear and the associated physical responses that occur when you hear a sudden, unexpected sound out of your line of sight show how these two systems are interrelated. Hearing is the sensory system that operates fast enough to underlie both.

Attention is about picking important information from the sensory clutter that the world (and your brain) throws at you twenty-four hours a day. At the simplest level, it is just the ability to focus on some events while ignoring others. But because attentional processes are, like hearing, continually ongoing, we are rarely aware of them unless we make ourselves aware of our behavior and have our attention drawn to our attention.

Do an experiment: go wash your hands. You've been sitting and reading for a while, and who knows where your hands or this book was a few hours ago. Before you get up, though, think for a second about the sound of washing your hands—you'll probably think of the water splashing in the sink and that's about it. But this time pay attention to *all* the sounds. The sound of your footsteps, whether shod or in slippers or socks, padding toward the sink. Did you walk on tile? Is your kitchen echoing with each footstep or are you wearing something soft and absorbent that damps it? When you reach for the faucet handle, do your clothes make a quiet shushing sound? Does the handle squeak a bit? What is the sound of the water as it is first leaving the faucet before it hits the sink? Is there a pattern to the sound of the water as it hits the metal or porcelain or plastic? If you don't have a stopper in the sink, does the water make a hollow resonant sound as it goes down the drain or does it build up and splash along the edges? What is the sound of water striking your hands as you move them under the flow? And when you turn the water off, did you notice how the flow stops masking the sound of the water draining away and the occasional drips of water from your hands?

That was maybe a thirty-second event. As with the sound walk I described earlier in the book, it took hundreds of words to describe the barest outline of the acoustics involved, events that you normally ignore but which involve the motion of trillions of atoms, the mechanical responses of tens of thousands of hair cells, and the activation of millions if not billions of neurons to register events that you normally don't even bother passing from sensory memory to consciousness. But there is a reason for this. If you paid equal attention to everything, with no automatic ability to parse out what was relevant to your needs, you would soon be overwhelmed by trivia both external and internal.

The mechanisms of attention have mostly been determined from studies using a technique called dichotic hearing, which simply means listening with two ears. An experiment in dichotic hearing goes something like this. Let's suppose you are presented with two different sounds, A and B, at different frequencies at different rates and times. If A and B are widely separated tones and presented randomly between your two ears, you tend to perceive them as random sounds. However, if you clump the two streams by some perceptual feature, such as their frequency separation (how far apart they are in pitch), their timbre (the fine structure of their sound), their relative loudness (quiet versus soft), or their location in space (left or right ear), your brain starts organizing them into separate acoustic streams. So if you are playing A-sharp on a clarinet every half second in the left ear and D-flat on a flute every three-quarters of a second in your right ear, you will group these stimuli into two separate auditory events—a very dull clarinet student on the left and an equally dull but distinguishable flute player on your right. But even if you start playing both sounds at the same time with the same rate, you will still separate them into different streams because your ears are differentiating not just the gross time of arrival of the events but the fine temporal structure of the sounds and their absolute frequency content. This type of study represents the simplest form of auditory scene analysis and one of the most basic measures of attention. As an experiment, it has stood up through time and changing techniques, with evidence ranging from early human and animal psychophysics and EEG through the most current studies using magnetoencephalograpy and fMRI.

But presenting individual tones varied by time, frequency, timbre, and position is a lab version of a caged hunt—it's not the kind of situation you run into in the real world. But it is the basis of a very common and robust effect called the "cocktail party effect." The idea is that even in a crowded room with lots of voices and background noise, you can still follow an individual voice (within limits, of course). The basis is again the synchronous responding to specific features of a complex sound even in noise that should mask it. It also highlights another aspect of auditory attention. In a room full of people speaking, you have a lot of overlap of sounds: a mix of male and female voices will have fundamental frequencies ranging from 100 to 500 Hz, some timbre features in common because they are all being generated from human vocal folds, different loudnesses depending on the distance from the listener and how loud they speak, and of course different locations. Amidst all of that, if you notice your significant other talking to someone, the familiarity of his or her specific voice and speech patterns (which drive the timing of the sounds made while talking) will activate auditory neural patterns that have been activated numerous times before.

Hearing something numerous times, even with the great variability in acoustic specifics that occurs in speech, not only causes auditory and higher-center neurons to respond synchronously but actually rewires the synapses in your auditory system to improve the efficiency of responding to those specific traits. This is a general form of neural learning called Hebbian plasticity. Neurons are not just simple connected structures that sum up input from earlier neurons and fire or suppress firing. Neurons are extraordinarily complicated biochemical factories, constantly synthesizing or breaking down neurotransmitters, growth hormones, enzymes, and receptors for certain brain chemicals, and constantly up- and downregulating them based on the demands placed on them during tasks. But the most remarkable thing about neurons is that they will actually grow new processes, so they can change their wiring pattern. Neurons that often fire in synchrony on a regular basis, such as those exposed to harmonic sounds commonly found in speech or music, will interconnect more so that they can more easily influence each other and work together. So something familiar in the midst of noise will jump out of the background by activating a specific population of cells that "recognize" this stimulus.

"Those that fire together wire together" is a general neural principle, but the difference between sound and vision is highlighted by comparing the cocktail party effect with the visual equivalent—a Where's Waldo? image puzzle. Trying to find the figure wearing the bright red-and-white striped hat and shirt in a complex visual scene usually takes at least thirty seconds (and sometimes much, much more; at times I could swear he is hiding in a cave in Afghanistan), whereas isolating a voice of interest in the cocktail party milieu usually takes place in well under a second.

But the cocktail party effect has an evil twin, one you are likely to run into. You're on a train or a bus, trying to read, sleep, or just not look at the guy across from you, but the person behind you keeps chattering into his cell phone. Whether he does it loudly or even just as a constant soft susurration, we still find listening to someone else's half of a conversation to be consistently annoying. A recent study by Lauren Emberson and colleagues found out why, and it has to do with the dark side of attention. They discovered that while hearing a normal conversation was not significantly distracting, hearing a half conversation—a "halversation," as they called it—caused a serious decrease in cognitive performance. Their hypothesis was that background monitoring of unpredictable sounds results in more distraction for a listener engaged in other tasks. Because you can't predict the direction of half a conversation, you get more unexpected stimuli, and thus more distraction.

This is one of the problems of having a sensory system that is always "on." Your auditory system is constantly monitoring the background for change. A sudden change in the sensory environment can break the brain's attention to a task and redirect it. Was that a footstep behind me in the dark alley, or just an echo from the wall across the street? Is that sudden howling a coyote looking for a neighborhood cat, or just the blind beagle down the block trying to get its owner to let it in? Sound is our alarm system twenty-four hours a day. It is the only sensory system that is still reliable even while we sleep (which probably served our ancestors very well when they were hoping not to be killed in their sleep by predators). A sudden noise tells us something *happened*—and the auditory system (as opposed to vision) operates quickly enough to provide sufficient synchronized input to determine whether the source of the sound is familiar or if we need to link it to additional sensory and attentional processes to let us reconstruct what that something might be even if we can't see it.

This is important because your other senses—vision, smell, taste, touch, and balance—are all limited in range and scope. Unless you turn your head (which has all kinds of consequences), binocular vision is limited to 120 degrees, with about another 60 degrees of peripheral vision. A smell has to be particularly concentrated for us to detect it at any distance, and even then humans have very limited ability to localize the source of a scent. Taste, touch, and balance are all limited to the extent of our bodies. But on a good day outside without any temperature inversions or earbuds blocking your ears, you can hear anything within a kilometer or so (about six-tenths of a mile)—if you are standing on solid ground (or floating on water), that's a hemisphere of about 260 million cubic meters or about the volume of 1,300 zeppelins, in case you measure things that way. But if you're standing outside studying for a test and your phone rings, your attention will swing to that, which is why the zeppelin lands on you—you're paying attention to the familiar if distracting ringtone and ignoring the more slowly encroaching shadow of the descending (and very quiet) zeppelin.

This scenario shows the difference between the two types of attention your ears and other senses have to contend with: *goal-directed attention* (listening closely to your phone conversation as you enter an area with spotty coverage) and *sensory-directed attention* (being unable to focus on your conversation because the man talking into his cell phone has said the word "bomb" three times in one minute while you're waiting for your flight to take off). Goal-driven attention makes us focus our sensory and cognitive abilities on a limited set of inputs and can be driven by any of our senses. Most of the time, humans are paying default attention to vision. You look around, reach for things by visual guidance (even if it's to turn up the volume), and get instantly irritated if the lights suddenly go out because some idiot installed infrared motion detectors to save electricity. But any sensory modality can serve for this type of attention. You can follow your nose to find the source of the awful smell coming from the bathroom. You can use your taste buds to keep altering a recipe until it tastes right. You can focus on putting one foot directly in front of the other and ignore the six-foot drop on either side of you if you are showing off by walking on top of a wall. And of course you can listen intently to try to figure out the guitar lick that has been killing your score in Guitar Hero. In all cases of this kind of attention, you highlight the input from the sensory modality (or modalities) that is providing the most information about the task you have consciously decided is the most important.

But stimulus-based attention is about grabbing and redirecting your attention from elsewhere, including goal-directed behaviors. It too can be driven by any sensory system—a sudden flicker in your peripheral vision, the smell of smoke, a sudden touch that makes you jump out of your skin. The redirection of attention based on a stimulus requires that something be novel and sudden—in other words, startling. A startle is the most basic form of attentional redirection, and it requires even less of your brain than the simplest auditory scene analysis. Being startled is something that happens to all of us: you're concentrating on something and then there's a sudden noise, or someone taps you on the shoulder, or (if you are in California) the ground trembles. And in less than 10 milliseconds (1/100 of a second), you do what I did when I heard the coyote-induced splash: you jump, your heart rate and blood pressure soar, you hunch your shoulders, and your attention swings as you try to find the source of the disturbance. You can be startled by sound, touch, or balance but not by vision, taste, or smell.* This is because these first three sensory systems are *mechanosensory* systems, relying on rapid mechanical opening of neurotransmitter channels that fire a very *fast,* evolutionarily very old neuronal circuit to activate spinal motor neurons and arousal circuits in your brain.

Every vertebrate has the startle circuit (with variations in expression), as it is a very adaptive way of putting an organism on guard for something novel. According to seminal work by Michael Davis and colleagues in rats (and subsequently confirmed in primates and humans), the mammalian auditory startle circuit makes its way through a five-neuron circuit, from the cochlea to the ventral cochlear nucleus to the nucleus of the lateral lemniscus to the reticular pontine nucleus and then down to spinal interneurons and finally the motor neurons. Five synapses in 1/100

---

* You can be scared by something moving in your peripheral vision, which is more responsive to motion than to form, but as it causes you to redirect your vision onto what startled you, it is much slower than an actual startle response.

second to go from a sudden sound to a sudden jump, putting you in a defensive posture with tightened, flight-ready muscles, sometimes emitting a loud vocalization (that's especially true of my wife). The auditory startle is common in vertebrates because it is a very successful evolutionary adaptation to an unseen event. It lets us get our bearings and get the hell out of there, or at least widen our attention to figure out what the noise was.

But being startled doesn't necessarily make you afraid. It does increase your sense of arousal—the physiological and psychological state that heightens everything from senses to emotional response. If you turn around and see that it's your friend who snuck up behind you and yelled "boo" while you were focused on the scary movie, then the sensation of fright goes away and what you feel is a sense of contentment, since she totally deserved getting your drink flung in her face from the reflexive raising of your arms as you turned your head. But what if you turn around and can't see anything around that might have said "boo"—no idiot friend with a peculiar sense of humor, no speakers in the ceiling, nothing? Then arousal continues, heightening everything, and emotion starts kicking in.

Emotions are a tricky subject to analyze in neuroscientific terms. Emotions are how you *feel*—complex internal behaviors that affect how you respond to what happens next in your environment until that emotion changes or fades away. Scientists (and others including philosophers, musicians, filmmakers, educators, politicians, parents, and advertisers) have been studying and applying emotional information for centuries. There have been hundreds of studies and books and even entire schools of thought about what emotions are, how they work, why we have them, and how we can use them, usually by manipulating them in others. Each of these treatments has conflicts with the others—

even lists of basic emotions often have little agreement, ranging from four pairs of "basic" emotions and their opposites, as proposed by Robert Plutchik in the 1980s (joy-sadness, trust-disgust, fear-anger, surprise-anticipation), through forty-eight separate emotions in ten categories, as proposed by the HUMAINE group, creators of the Emotion Annotation and Representation Language. Given that there are disputes even about the basics of what emotions are and what causes them (whether physiological states, such as the James-Lange theory suggests, or cognitive ones, as proposed by Richard Lazarus), it is not surprising that attempts to identify the neurobiological underpinnings of emotions are contentious. But one thing that is consistent in studies of emotion using techniques ranging from nineteenth-century psychology through twenty-first-century neural imaging is that one of the most important and fastest-acting triggers for emotion is sound, distributed throughout the cortex by both tonotopic and non-tonotopic pathways from the medial geniculate. So how does sound trigger and contribute to specific emotional states?

Fear is one of the most studied emotional states, possibly because it is sort of an emotional primitive—even frogs can feel fear (if you hang out with them long enough to understand their highly non-verbal body language). It is also one of the few emotions with a well-characterized anatomical and physiological basis, and it is usually studied with sound using a very old technique called classical or Pavlovian conditioning. The basis of classical conditioning is relatively simple: you have an unconditioned stimulus, something to which you would have a reflexive response, such as the sight of a steak making a hungry dog (or grad student) salivate. The steak is the unconditioned stimulus and the salivation is the unconditioned response. The trick Ivan Pavlov came up with is stimulus substitution. Right before showing the hungry dog the steak, he rang a bell (the conditioned stimulus). After a few repetitions in which the experimenter rang the bell within half a second of presenting the steak, the dog's brain created an association between the ringing of the bell and the presence of the tasty steak and actually rewired its reflexive response to salivate in response to the bell, even if there was no steak.*

Scientists who study fear rely on classical conditioning and other techniques using sound, in large part because there are traceable neural pathways leading from the ears to a region of the brain that modulates fear and fear-like responses, the amygdala. The amygdala is one of those rare brain nuclei that gets a lot of press—though most of the information out there in the popular media is wrong. If you get your science from news websites, you will often hear about new and exciting findings showing that anything having to do with control of emotion or the failure of such control—criminal behavior, political orientation, the perfect diet, the ultimate date, why chocolate is the same as sex—is associated with the amygdala somehow. The popular press view is that the amygdala is the brain's "emotional center" and that if you do something socially unacceptable, it's because your amygdala is screwed up.

The truth about the amygdala is much more interesting. It gets input from both the fast (thalamic) and slow (cortical) pathways and provides output throughout the cortex. The fast pathway provides a quick and immediate reaction as well as a basis for learning the difference between dangerous and nondangerous input; the cortical pathway, which wends its way

---

* At least for a while. There are entire psychological disciplines and thousands of books and papers on the strategies to get maximum conditioning to a conditioned response.

through the memory and associational regions of the brain, is slower but much more accurate in determining if an emotional response to a sound is valid. Was that a scream in the film you're watching or was someone just butchered behind you? (A really good film will blur that boundary because of good sound editing; we'll talk more about that later.) But a real and complex fear response is triggered rapidly, then maintained and strengthened due to a network of connections in your brain, not just the amygdala. Sounds associated with fearful events get looped forward and backward through the amygdala, the auditory cortex, and the hippocampus, the last of these being the gateway to long-term memory storage. Early in these loops, information is fed to deep cortical regions and hypothalamic areas associated with control of autonomic systems such as blood pressure and heart rate, modulating the rapid release of adrenaline, making your heart beat faster, constricting your pupils, and drying out your mouth. A frightening sound, such as a very loud lion's roar, causes body-wide physiological responses that feed back to your brain, building on the early emotional response and simultaneously comparing older memories of frightening events with new input to see if what frightened you is still really scary. If the roaring sound has faded, meaning you outran the lion, your sympathetic nervous response starts to diminish and you calm down, still aroused and wary but with the time to look around and confirm that you are no longer in danger.

Certain elements of sounds act as basic triggers, even without previous experience. Sudden loud sounds can cause you to startle, but if you add in very low pitch, your brain starts making subconscious associations. As with frogs selecting mates, loud and low-pitched means large. Large sources of sounds might be desirable if you are a female frog looking for a mate, but in a normal human day, large is often scary. One evolutionary argument has been made that we have been biologically hardwired to interpret loud sounds at the very low end of human hearing and lower, in the realm of infrasound, as signaling "predator." Analyses of the roars of big cats such as lions and tigers have demonstrated the presence of a high-amplitude infrasonic element. Evolutionary neuroethologists, who study the ways our brain-driven behaviors have changed through evolution, have argued that those animals that did not automatically run away when they heard such sounds were the ones that got eaten and didn't pass on their genes. But another hypothesis, which combines non-auditory physiological acoustics with a tie-in to autonomic sensitivity, points out that loud infrasonic sound not only is heard but is felt throughout the body, vibrating the organ-filled abdomen, the air-filled lungs, and even the bones themselves, similar to the non-tympanic pathway in frogs. Vibrating the rest of the body at low frequencies can induce nausea and a feeling of sickness via the enteric nervous system, a poorly understood subsection of the autonomic nervous system that provides sensory feedback from your internal organs. This low-frequency non-auditory pathway underlies things such as vibroacoustic disease, which affects some construction workers with excessive exposure to jackhammers and other construction-based vibration. So a sudden, loud low-frequency sound is not merely triggering basic auditory connections but taking input from your whole body and telling you to *run*. Interactions with memory and attention regions come later (and more slowly) but guarantee that you will remember that the sound was associated with danger, and help you respond even faster next time you hear it, thanks to Hebbian plasticity, the brain's ability to rewire itself to speed up responses to previously encountered events. The scary sound has become a survival tool—next time you hear that roaring sound under the same circumstances, you'll probably start running faster and not waste time yelling, "What the hell was that?"

What about sudden loud sounds that are not particularly low-pitched and hence probably not life-threatening? Then the context provided by memory and previous associations becomes more important. For example, you're cruising the web and you hit a web page that suddenly screams "Congratulations— you just won an iPad" or blares some band's latest thirty-second low-sample-rate offering. Irritation closely follows the startle, and you almost instantly close that page (or take the more drastic measure of turning off sound when browsing the web). Sounds that are intrusive but not immediately associated with danger assume what is called "negative valence"—feelings like annoyance or anger. They are a response to a false alarm: you got all ramped up because of a sudden sound, but it's just another familiar irritation. This is one reason the use of complex sound in technology is problematic. Remember the talking-car warnings of the 1980s? A sudden poorly synthesized human voice with no one in sight telling you "Your door is ajar" or "Your seatbelt is not fastened" was startling and rapidly categorized as irritating, especially when it would preempt the stereo.* More recently, most web pages (at least those from smart designers) no longer use complex sound as soon as you land on them—the sudden sound is an irritant, and if it lasts longer than the few

---

* A friend of mine actually did a thriving pirate car service in killing the voice synthesizer chips in these cars, apparently the most common and most irritating being those in Chrysler LeBarons.

seconds it takes for you to visually recognize the image as something you find interesting or desirable, you close the window.

One of the most widely known negative-valence sounds is one you probably tortured classmates with back in elementary school. Back in 1986, Lynn Halpern, Randolph Blake, and James Hillenbrand wrote a wonderful paper called "Psychoacoustics of a Chilling Sound" that did something that far too few scientific papers do: it addressed a really basic question. Why do we hate the sound of fingernails on a blackboard? (Not to mention metal, such as a rake, being dragged across slate, which was rated as an even worse sound.) They hypothesized that the spectrum of the sound closely matched that of the warning cries of the macaque monkey, and so fingernails on a blackboard was the sensorineural equivalent of a primate alarm call. This was a cool idea, and it was widely cited, but, as with most non-mainstream scientific studies, it didn't get much further testing beyond a paper by Josh McDermott and Marc Hauser in 2004 showing that cotton-top tamarins (very small, very cute monkeys) didn't show the same response, possibly due to the fact that they rarely interact with blackboards or metal rakes. So the "Chilling Sound" study's results remained unconfirmed. But the psychological effect of this sound is nearly universal among humans—fingernails on a blackboard and metal on concrete both create sounds that make you want to plug your ears and throw sharp objects at the person making the noise, no matter your age, sex, occupation, or cultural upbringing.

Having lots of sound equipment about and being an academic, I had access to blackboards and fingernails, as well as a metal rake and a concrete driveway. So, after putting in earplugs and sampling both these sounds, I noticed something very interesting: the fine timing structure of both sounds was actually what's termed "pseudo-random"—that is, the underlying waveforms were *almost* periodic, repeating their fine structure over time, but there was enough random variation to make it a temporally messy sound. It's sort of like looking at a badly Photoshopped image, with the ragged edges, out-of-place colors, and pixels making it obvious that something is wrong. The only other time I'd seen anything like that was in a recording of people shrieking at the top of their lungs. And then it snapped together in my head: our reaction to these sounds is probably based not on the frequency content of some ancestral warning or alarm call but on the pseudo-random variations in the fine time structure, just like those that show up when someone screeches uncontrollably as if in pain or a panic and the normally harmonic structure of the voice becomes ragged and out of control. It isn't the whole sound that causes the reaction, just a piece of it, but such a basic piece that it elicits a profound and unpleasant reaction.

So what does this mean? If we hate pseudo-random sounds, do we like regular periodic ones? Well, sometimes. As mentioned earlier, living things tend to make harmonic sounds, with regular mathematical relationships between frequency bands and very periodic timing. But sometimes a fine acoustic detail can shift the valence of a sound, turning our associations with it either negative or positive. One of my favorite demos is one I discovered by accident when playing with a sound editing program. I ask students to rate two sounds for various factors, basically ranging from comforting to alarming. I start off by playing a sound that is almost universally frightening: the sound of angry bees. Even though it's not low-pitched or even necessarily very loud, this sound is definitely fear-inducing, and not just in humans—one study showed that elephants will actually move far away and emit specific warning vocalizations when they hear this sound. My students all rate this as an alarming sound. Then I play a sound that is almost universally calming: the sound of a cat purring. This is almost always rated as a pleasant sound. But then I add a twist: when I take the sample of the cat purr and speed up the amplitude modulation rate from a few cycles per second to several hundred per second, it begins to sound just like alarming bees. Taking the calming respiratory-like sound and speeding up its repetition rate (and throwing in a little randomness) changes the valence of the sound radically, mostly likely based on previous experience and associations of the two sounds.★ In another demo that shows how powerful previous associations can be, I put on a sound that has sort of a sizzling, ratchety repetitive quality. On half a sliding blackboard, the sound is identified as "rain on a sidewalk." I ask students to rate it: is it a nice sound, is it pleasing, does it make them comfortable or uncomfortable? Most of them say it's soothing, reminding them of rainy afternoons with not much going on. And then I shift the board over to reveal the line "it was actually mealworms devouring a bat carcass." You can almost see neurons melting in their brains as they go from neutral to "ewwwwwwww" in three-tenths of a second.

But what about a lack of sound? Silence, especially in a normally noisy context, can be an extraordinarily powerful emotional acoustic event. Since we are always presented with subconsciously monitored background noise, a sudden lack of

_____

★ But therein lies the potential problem of relying on "universal" emotional associations. There was one student who absolutely shuddered when she heard the cat purr. After class I asked her why she seemed so uncomfortable. She replied that she loathed cats and anything associated with them. Her personal history had changed her response patterns to a sound most others found very agreeable.

outside sound leaves an awful lot of attentional and arousal control bandwidth available. A perfect example is the old movie cliché: two explorers are wandering through a jungle when one stops and says, "Do you hear that?" The other responds, "I don't hear anything," to which the first one responds, "Exactly. It's too quiet." The detection of the absence of sound, while slower than the detection of a sound, triggers its own set of responses, increasing attention and arousal, which can lead to internal mechanisms of increasing your ear's gain or sensitivity. This increased arousal from silence has the same effect as increased arousal from a startle or a frightening sound—it heightens your emotional preparedness. A study by Denis Paré and Dawn Collins examining conditioned responses to a series of tones followed by a silent period showed increased blood pressure and synchronization of cells building up steadily during the silent period. This suggests that silent anticipation before something unpleasant was critical for learning about unpleasant or scary stimuli. It may be that inappropriate silence is not so much frightening in itself but sends signals throughout your brain that something is missing, something is *wrong*, preparing you for something bad. It may be as basic as the lack of crickets in a forest at night, making your hindbrain wonder if they heard something padding about that you didn't, or something as complex as the utter silence of UC Davis students as the school president walked by after an incident in which a police officer pepper-sprayed some non-violent protesters on campus—a social warning by denial of sound.

Sound can be very effective at generating a negative response, but the world is full of sounds that evoke other emotions. If you've ever been to New York's Penn Station, near where the Long Island Rail Road waiting area is, you know that it is not a particularly pretty environment. But one day in the dead of winter when I was waiting for the eternally late train, I suddenly heard birds chirping and singing. My first thought was that some local robins had flown into the station and were singing happily because they were not freezing to death in Central Park. But upon looking around and seeing no birds, I spotted a well-hidden speaker. It was playing a long, looped recording that sounded like a natural, if sparse, birdsong chorus. It was a genius piece of emotional engineering—play sounds associated with the countryside and spring mornings while people are surrounded by urine-drenched pillars and waiting for a train that will no doubt be several years late, and you will take the edge off the tension of hundreds of stressed-out commuters.★

The basis for positive emotions, whether triggered by sound or anything else, is more complicated to understand. There is no simple anatomical region that underlies positive or complex emotions. This is probably because fear and other negative emotions drive survival behaviors such as running away or fighting, while more complex ones come with developmental, behavioral, and cultural baggage. What is erotic or intriguing or relaxing to me might be dull or horrifying for you. There have been a lot of attempts at modeling complex emotions such as love using animal models—after all, it's very difficult to get an experimental protocol past a review committee by saying, "We're going to drop electrodes into people's brains to identify love areas and inject neural tracers, then slice up their brains to see where love lies." Scientists have examined maternal care for offspring in everything from rodents to sea lions, pair bonding

★ Unless, of course, they had been previously mugged by a bunch of tough bluebirds.

in prairie voles (who seem to release the same neurotransmitter, oxytocin, as humans involved in positive social interactions), even mourning behavior in elephants and primates.

But in any of these comparative studies, it's hard to jump the species barrier for anything more complex than fear, anger, or other reactive emotions—we simply don't know whether other species experience more subtle feelings. Is a mouse happy when it hears the bell sound that means a treat is forthcoming? Or is it a simple association with a forthcoming reward? The closest neuroscience has come to identifying the sources of complex and positive emotions is identification of reward pathways and systems, such as the septal nuclei and the nucleus accumbens. Both structures are highly involved in pleasure-seeking behaviors. Back in 1954 James Olds and Peter Milner demonstrated that if you implanted stimulating electrodes in the septal nuclei, a rat would continuously stimulate itself, even ignoring food and water. The nucleus accumbens is highly responsive to drugs such as cocaine and heroin and is seen as an important site for generating the reward in satisfying addictive behaviors. Both of these sites are intricately connected with all the deep cortical structures that underlie subconscious motivation. The nucleus accumbens also receives a great deal of dopaminergic input from the ventral tegmental area, a region that connects bidirectionally to all areas of the brain, including those in the deep brain stem that connect to the cochlear nucleus. The presence of a relatively fast deep connective route from the basic auditory nuclei combined with massive inputs from all other regions of the cortex is similar to the arrangement found in the amygdala, and a study in 2005 showed that the nucleus accumbens plays a role in changes in emotional state induced by music.

But therein lies the problem. Neural imaging allows you to

look into the living human brain without doing things that will get you prosecuted under the Geneva Conventions. Much of the current crop of neuroscience research looking at positive emotions arises from neural imaging studies using fMRI, which shows where there is an increase in blood flow in specific brain regions based on presented stimuli. A great many studies have been run that claim to examine the underlying basis of things such as love and attachment, but all too often these claims end up blowing up in the claimant's face because, to put it simply, complex stimuli elicit complex reactions, and fMRI is a crude tool for measuring them. One of the more glaring examples was reported not in a scientific paper but in a *New York Times* op-ed piece by Martin Lindstrom, a well-known consultant who has done some interesting work in the field of neuroeconomics, the study of how we decide what to buy. The study he reported on involved examining the response of young men and women to the sound or image of a ringing and vibrating iPhone, and the claim was made that subjects showed activation in both visual and auditory cortexes regardless of whether they had heard the iPhone or seen it. This was intended as proof that they were undergoing multisensory integration. The piece went on to claim that because the most activity was seen in the insular cortex, a region that some studies have shown is associated with positive emotion, that the subjects *loved* their iPhone.

This is the type of claim that should make us question neural imaging studies, especially those applied to commercial interests, and in fact it inspired more than forty scientists to write a reply to the paper highlighting the significant problems with this study. First of all, identifying the insular cortex as being "associated with feelings of love and compassion" is not useful, as this region of the brain gets activated in about one-third of *all* neuroimaging studies. Second, the insular cortex—like most other parts of the brain—is involved in a lot of brain-directed activities, ranging from controlling heart rate and blood pressure to telling if your stomach or bladder is full. In fact, the insular cortex is involved in almost every inward-oriented process we go through. So identifying it as *the* place that makes you love your iPhone made for a great marketing moment but a very poor scientific claim.

As much as we neuroscientists love our tract tracing, EEGs, fMRIs, and PET scans, the complex mind is still much of a black box. When we want to determine what our subjects feel, we sometimes have to fall back on a good old-fashioned technique: asking them. Gathering subject response by questionnaire is even more top-down than running someone through an fMRI machine, but it gives you a handle on the actual cognitive or emotional response rather than worrying if you are identifying the right chunk of brain. Still, questionnaires are fraught with their own limitations. Questions have to be carefully constructed to avoid biasing the subjects. Subjects may answer based on their perception of how they think they are supposed to answer rather than how they actually feel. The environment they are being tested in, a lab or a classroom, may have its own emotional associations for them, so their answers are altered by their environmental context. Sometimes what they actually feel can't be described by a simple linear scale from "pleasant" to "unpleasant," or "arousing" to "calming." On the other hand, the good thing about this technique is that you can pay a *lot* of volunteers to give responses for what it would cost to run even one fMRI scan, and large numbers provide a lot of statistical power when trying to answer complex questions about the mind. (Plus these statistics then form the basis for running the more expensive tests.)

Not surprisingly, sounds are among the most common and powerful stimuli for emotions. There are a number of standardized psychological databases that assign emotional valence to different non-verbal sounds.* These databases have been used for decades and have accrued so much data that they are used as the foundation for studies employing other, more technologically oriented techniques, from EEG to fMRI. But even with this basic a technique, you run into the problem of trying to apply operational definitions to subconscious responses. Imagine you are a fifteenth-century barber/alchemist/wizard who is really interested in categorizing emotional responses to sounds. Lacking a digital recording device, you have your subjects sit in a dark room with a curtained window while you make sounds out of their view. They consider the sounds of crashing armor from a joust to be highly arousing and pleasurable, whereas the low grunting of a wild boar is frightening. Then play these sounds for someone in the twenty-first century (who does not spend a lot of time watching *Game of Thrones*). The crashing armor would probably be considered arousing but annoying, whereas the wild boar call becomes the sound of just some animal rather than the thing that killed half your village. For example, in one database, the sound that was rated as most pleasurable was a ten-second sample of the intro to Joan Jett's "I

---

* It should be noted, though, that a lot of these databases have their own baggage, since their data, as with most research done on humans, were drawn from American college students who really needed the extra $20 or the extra credit.

Love Rock n' Roll." This seems like an odd choice for most pleasurable sound—except the database came out in the late 1990s and the song was a big hit in 1981, when the people who were the sources of the data largely were impressionable pre-teens with access to MTV. It's pretty unlikely that this particular sound effect will stand the test of time, but it's pretty likely that one of the ones rated most unpleasant—the sound of a child being slapped and then crying—will continue to be rated the same way.

Sounds that are familiar will be processed faster, and those that involve things important to humans in general, such as hearing one of our young being made unhappy, will generate a stronger emotional response even if you have been tempted to slap a kid yourself after listening to it crying on a plane for six hours. However, another study by Melanie Aeschlimann and colleagues in 2008 pointed out that using a wide variety of sounds of different lengths and loudness could introduce too many processing artifacts. Since you are able to respond to sounds in hundredths of a second or less, a ten-second-long sample could cause a listener to only respond to the last few seconds or respond on some sort of internal summing mechanism. Perhaps the rating "most unpleasant" was based not on hearing the sound as child abuse but rather on a previous experience of a child's sustained crying. This study proposed a completely different database based on samples two seconds long with a somewhat different rating metric. The samples were not complex sounds or speech, but rather human non-linguistic sounds (screams, laughs, erotic sounds) or non-human sounds such as alarm clocks. Using these very short samples, the researchers found that there was a lot of overlap with ratings of longer samples, indicating that the emotional valence kicks in very rapidly. However, a few interesting things popped up. First, sounds that had negative emotional responses tended to be perceived as louder even though they were at the same amplitude. Second, the strongest ratings were associated with those sounds labeled as emotionally positive. And finally, the sounds with the strongest emotional response in any category were human vocalizations.

Sounds that evoke the strongest emotional response tend to be those from living things, especially other humans. Mechanical or environmental sounds tend to grab your attention but are usually limited in how far they will take your emotional response, unless they tell you of specific dangers you want to confirm visually (such as the sound of a rockslide) or unless they have strong associations (such as the sounds of waves at a beach). Sounds deliberately made by living things are almost always communicative—the dog's growl, the frog's croak, the baby's cry—and are almost always harmonic (with variations from harmonicity bringing their own emotional response, such as cringing when someone screams).

The frequencies that make up human sounds (including but not limited to speech) and sounds from animals about our own size are in our most sensitive region, so it is easier for us to hear these sounds—they will jump out of the background. In addition, sounds that we have heard before are more easily identified and reacted to more quickly. Add on top of that the fact that we process low-level sensory information such as changes in tone and loudness faster than we do complex input such as speech, and we begin to understand why we can react quickly to the emotional content of sound. You don't need to have been bitten by a dog to know that a low growl is menacing or have been mugged by a squirrel to know that a harsh barking sound means you're too close to its territory. Overall, *any* communication is about first evoking an emotional response on the part of the listener; humans just glue semantic content on top of it, giving us a tenuous hold on the title of most intelligent beings on Earth.

I will not delve into the intricacies of human speech (and thus triple the size of this book), but I will note that the emotional basis of communication relies not on *what* is said but on the acoustics of *how* we say something, independent of the formants of speech and to some extent what language is spoken. This flow of tones, called *prosody*, was first discussed by Charles Darwin as a predecessor of human speech, and to some degree the idea still holds up today. Both neural imaging and EEG studies have demonstrated that prosody is processed not in Wernicke's region of the brain (which underlies speech comprehension) but rather in the right hemisphere. This is the side opposite to where the linguistic processing centers are in most people but in the region more important for contextual, spatial, and emotional processing.

For a simple demonstration, just say the word "yes." Now say it as if you had just found out you won the lottery. Now say it as if someone had just asked you a question about something in your past that you thought no one knew about. Now say it as if this is the fortieth time you've answered "yes" in a really boring human resources interview about how much you love your job. Lastly, say it as if you were just forced to agree to a really horrible contract in order to keep from losing your job. Linguistically, you indicated affirmation every time, but each time the emotional meaning differed. What you changed was *how* you said the words: overall pitch, loudness, and timing. And someone listening to you who was also a native speaker of your language would get the subtext, which is sometimes even more important than the ostensible meaning of the utterance. For example, think about how hard it can be to understand some synthesized speech patterns, particularly older ones. Speech synthesizers from the 1980s until the late 1990s were pretty much playbacks of recorded phonemes with no reliance on prosodic flow—in other words, the speech sounded like it was coming from a robot. Even today, after millions of dollars poured into making synthetic speech sound more "human," you can still hear the difference between a human voice and, say, the iPhone "operator" Siri.

The lack of normal emotional undertone in a robotic voice not only makes it harder to relate to but also contributes to the annoyance most people feel when stuck dealing with one, even if it does provide the same information a human being would. On the other hand, think about the immediate comprehensibility of the non-verbal sounds coming from a favorite movie character, R2-D2 from *Star Wars*. R2D2ese was created by legendary sound designer Ben Burtt using filtered baby vocalizations for some sounds and completely synthesized boops, beeps, and arpeggios for others. Whether the fictional droid was excited, upset, happy, or sad, the audience had no doubt what was going on in its CPU even though the sounds were completely devoid of linguistic content and formed entirely from prosodic tone structure and context. But prosodic emotional comprehension is a *part* of human language and communication and is at least partially language-specific (R2D2's vocalizations were in fact generated by English-speaking people, although some of my native-Russian-speaking friends claim they had no problem understanding what was meant).

This is why prosody is not a universal language—the bird chirping that makes you feel so relaxed because it reminds you of a spring morning may in fact be coming from a robin who is really pissed off that a bunch of cowbirds have invaded her nest. But it raises some interesting evolutionary questions regarding human communication. Was prosodic sound a precursor to the first human language? There are some universal elements common to all sonically communicating species, such as loudness for emphasis, lower pitch for implication of size or dominance, faster tempo for urgency. And despite linguists' habit of dividing languages into the categories of tonal (such as Mandarin) and non-tonal (such as English), even the basic structure of words, phonemes, shares common acoustic elements between languages based on the underlying biology of how we make sounds. A recent study by Deborah Ross and colleagues showed that if you examined the frequency distribution of phonemes uttered by male and female speakers in both Mandarin and English, similar patterns emerge. Whether in a tonal language or a non-tonal one, speech sounds most commonly organize around twelve tone or chromatic interval ratios—more commonly known as the twelve steps in a musical scale. The very basis of even non-prosodic human language has its roots in the mathematical underpinnings of sound. The question that arises is, how do these mathematical relationships tie together sensation, emotion, and communication? For that we turn to one of the hardest things science has ever had to face: music.

Chapter 6

TEN DOLLARS TO THE FIRST
PERSON WHO CAN DEFINE "MUSIC"
(AND GET A MUSICIAN,
A PSYCHOLOGIST, A COMPOSER,
A NEUROSCIENTIST, AND SOMEONE
LISTENING TO AN IPOD TO
AGREE . . .)

ABOUT THE TIME I started working on this book, I was contacted by Brad Lisle of Foxfire Interactive to see if I would be willing to be a science consultant for a 3-D IMAX film about sound, titled *Just Listen*. The very idea of this blew me away—how do you take such an immersively visual medium as a 3-D IMAX film and make it focus on something as non-visual as sound? But Brad has remarkable ideas about education and interaction. He wants to teach people about the sonic world they are immersed in by providing a medium that would refocus their attention. His initial interest in me was due to some of my work in visualizing how bats perceive the world. I found that if you created an animated world with virtual objects made of crystal-like glass material and swapped out the usual lighting parameters for sonic ones, you could create a world of acoustic glints that formed recognizable 3D shapes as the viewer and objects moved in it. This gave us a handle on how bats perceive the world with their ears, translated into the more common human motif of vision.

But as much as I love them and as common as they are, bats are not really something that most people think about. And most people who study sound tend to take very mathematical physics-based or psychological-neuroscience approaches. Most audiences who go to even a science-based IMAX film are not really there to learn about near-field effects or frequency-dependent acoustic spaces; they are there to have an experience. Among all the elements Brad and I discussed for the movie, ranging from animal communication to the sounds of human spaces, something was needed that would tie everything together as soon as the movie started. And of course, the one thing that could tie it all together, the one type of math that everyone gets, the one complex neural phenomenon that we all appreciate, is music. But the film couldn't use just any music—both the music and musician had to be not just spectacular but also able to teach something about the very nature of music and the mind. So in August 2011 my wife and I went to Vancouver to work with the *Just Listen* team in recording a remarkable musician, Dame Evelyn Glennie.

For those of you unfamiliar with Glennie's work, you've missed something truly remarkable and beautiful. Glennie is a world-renowned percussionist and the only solo percussionist of the last century. When most people think about percussion, they tend to think about it either as just providing the beat for real music or in terms of the aggravation and headache that usually follow when you are stuck in a rock concert for the obligatory ten-minute drum solo that everyone but the drummer's mother could have lived without. But Glennie owns and plays almost two thousand different instruments, from recognizable classics such as marimbas and xylophones to custom-made and haunting oddities such as the waterphone. What I was fascinated by as I watched her play a short piece she had written was the fact that she wasn't playing the instrument so much as she was playing the room itself. Walking barefoot up to the six-foot-long concert marimba, she positioned herself and the instrument carefully, then rose up on her toes, tilted her head back, and with four mallets struck the first notes and made the whole room *ring*.

Seated curled up on the edge of the stage, trying to avoid knocking over nearby equipment, I felt the stage tremble and walls of sound fill the space and bounce back like a tidal wave toward the source. In the few seconds before her next notes, everyone could hear the strings of the unplayed grand piano positioned behind her resonating with the force of the near-field sounds from her strike. It was as if she had created a sonic sculpture that changed over the course of the first few seconds. It reminded me of what a fellow band member had said to me years ago: "You can't record music; you can only save CliffsNotes of it. You have to be in the room with it or you're just using it to fill the empty spaces between your ears." As Dame Glennie launched into the remainder of the piece, using her whole body to play the instrument, but always with her bare feet in solid contact with the floor, her head thrown back, exposing her neck and body to the vibrations from the marimba, I realized that I was in the presence of someone who personified the complexities that science has with dealing with music. Evelyn Glennie was filling a space with music.

Oh, and by the way: Glennie is mostly deaf.

Literally thousands of years of research into sound and music, starting with Pythagoras's study of the mathematics of musical intervals through the most current fMRI neural imaging study, treats music as an auditory phenomenon. Neuroscientific and psychological studies examine how music is detected by the ears, perceived and processed by the brain, and finally responded to by the mind. So if all that is true, how does Dame Glennie not only create music but listen to it? Because her definition of music and sound is different from anything you may read in a scientific paper. When I asked her a simple question, one that is capable of starting fistfights at scientific conferences—"What is music?"—her reply was that music was something that you create and listen to with your whole body, not just through your ears.

Being the science consultant in this type of situation gives you a certain leeway. While everyone else was positioning 3-D cameras, hoisting microphones, adjusting power supplies, and generally making sure that the next recording would be just so, I set up my own equipment. I knew that Glennie's deafness was not complete but that she had very little high-frequency information, so I set up an experiment. I placed two regular PZM boundary microphones on the stage out of range of the camera and where she wouldn't step on them. These types of microphones are often used to pick up the sound of the whole stage area during live recordings and can pick up sounds to about 22,000 Hz, slightly beyond the upper range of human hearing. This may seem like overkill, since the range of most musical instruments tops out at about 4,000 Hz (which is, interestingly, also the upper limit for human auditory hair cells to phase lock), but as anyone who has listened to music through cheap, blown-out speakers will tell you, without those high-frequency components, music sounds dead. Next to them, I placed two geophones, seismic microphones that only pick up low-frequency

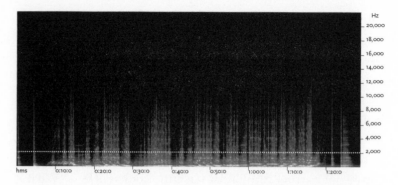

(Figure A) Sound recording of marimba piece. Spectrogram showing the frequency response (vertical axis) over time (horizontal axis) of the PZM microphones over a range from 0 to 22 kHz. Horizontal dotted line shows the 2 kHz cutoff range from the geophone recordings.

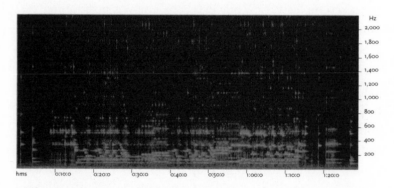

(Figure B) Seismic recording of the same marimba piece. Spectrogram showing the frequency response over time for the geophones, with a cutoff at 2 kHz.

sound and vibration, with the highest senstivity topping out at a few hundred cycles per second—approximately the range of Dame Glennie's hearing sensitivity—though they also pick up sounds far below human hearing in the infrasonic range. By feeding all of these into a portable digital recorder, I could record both the sound range heard by people with normal hearing and a simultaneous version restricted to low-frequency impacts and vibrations, similar to what Glennie heard.

Looking at the figures above, you can see the difference between the same piece of music recorded using a microphone that mimics the pickup of the human ears (Figure A) and the same recording done with the seismic microphones that are only showing distortion from interacting vibrations above a few hundred Hz (Figure B). While in Figure A you can make out the harmonic structure of the music within the regular vertical banding and the tempo of the piece through the brightly colored regions of the mallet strikes, you can see that there is a "cloud" of reverberation around each note played, formed by the reverberations from the individual notes filling the room, bouncing back and changing. This is the level of complexity that we hear with live music or well-done studio post-processing. But if you look at the spectrogram of the geophone recordings (Figure B), you can spot the individual mallet strikes as geometrical shapes with clean spaces around them, going only up a tenth as high as the acoustic recording. You can almost see the musical score in the percussive strikes.* If you compare the two types of recordings, the first one is obviously the acoustically rich one, the one our brains have evolved to perceive, decode, and translate into emotional responses, the one we would call "musical." Yet the lower one is closer to what the source, Dame Glennie the musician, actually perceives. While most of us sit there listening with our ears, perhaps (like me) with our eyes

---

* In fact, for those of you who have never had to take music lessons, musical notation is a series of dots or blocks placed on a vertical scale that represents the sounds in a spatiotemporal pattern, rather like a seismic spectrogram.

closed, she is picking up vibrations from the stage through her bare feet, near-field waves of sound from the resonators striking her legs and lower body and her neck, feeling the feedback from her hands and arms up through her body, some of it even resonating her skull, providing low-frequency sound through direct bone conduction. Glennie is using her whole body to detect vibration, and this is what she—accurately—calls music. Right here is the heart of the conflict that has always existed between science and music: the question "What is music?"

One of the biggest problems facing scientists who study music lies at the very heart of science itself. Science is about observing phenomena, questioning them, forming and testing hypotheses, and following up those that are successful and (if we're honest and not too politically funded) dumping those that are not. But at the start of any hypothesis is the need to form an operational definition. What are you testing? What are you studying? What parameters can you change without changing what you're studying? Music is notoriously elusive in this way. The title of this chapter is actually derived from how I start my lectures on music and the brain, and even though it sounds like I'm being a smartass, I'm not. I've been a musician, composer, sound designer, producer, and scientist and even I have trouble coming up with a definition that will satisfy more then two or three parts of my past and present.

Science has been trying to tackle music for centuries. In the last thirty years, I've probably read forty books and several hundred papers on the relationships between science and music; the biological underpinning of music in the ear, the brain stem, and the cortex; psychological bases of music perception; and musical cognition and its relationship to intelligence and the mind.

WHO CAN DEFINE MUSIC

These range from Christian Schubart's early nineteenth-century silliness that gave jargon-laden descriptions of the emotional bases of different keys through Helmholtz's classical work on the perception of tone.* Studies have been carried out by instrumentation as simple as glass bowls that would resonate at specific frequencies and by technology as recent as neural imaging, a favorite of contemporary theorists. Even archaeologists have gotten into the arena with the discovery of 35,000-year-old flutes made of bird bones, pushing the human relationship with music back toward the Cro-Magnon era.

But to tackle music scientifically, we need a good operational definition of what it is. I've seen several dozen attempts at it, ranging from the dictionary-like "a series of tones arranged in a precise temporal structure" (which would probably leave out all ambient, hard-core jazz and a huge amount of non-Western music) to the cognitive "an acoustico-emotive communication form," which winds up including things such as birdsong and gorilla chest thumping.

Music *seems* like it should be amenable to scientific analysis. It's composed of tones in an orderly (or deliberately disordered and hence non-random) temporal arrangement. It exercises control over frequency, amplitude, and time, three elements we've been throwing analytical tools for centuries. But once you've actually picked up an instrument and gotten it enough under control to elicit a response from an audience, even if it's just

---

* Schubart's *Ideen zu einer Aesthetik der Tonkunst* (1806) can be found at several sites on the Web, including www.gradfree.com/kevin/some_theory_on_musical_keys.htm. It is in vague agreement with Nigel Tufnel's claim that "D minor is the saddest key," although calling it "melancholy womanliness, the spleen and humors brood" is pushing it a bit.

yourself, you realize that music isn't the notes, the timing, the buildup and relaxation of psychological tension between all of those elements, or even how the player or listener feel before, during, and after.

Still, music requires all of these things, so we're kind of stuck. Music is a global and subjective subject. Just ask your parents how they feel about the "noise" that you like listening to or your kids why they listen to "that crap." And science has trouble with subjectivity. Without precision and testable definitions, you end up with statements like "D minor is the saddest key,"★ with hundreds of classical musicians nodding in agreement as they think about Bach's Toccata and Fugue or Beethoven's Ninth, and an equal number of neuroscientists and psychologists wondering why they can't see that in their fMRI data. This is what I find fascinating about trying to figure out the relationship between music and the mind. Their mutual complexities are almost mirror images of each other, and in understanding one we may discover the underlying basis for the other. But none of the hundreds of books or thousands of studies carried out has truly grasped the interface between the two. Perhaps the best we can do is to treat it like a complex jigsaw puzzle, looking at small pieces along the edges and seeing how they fit to create a basic framework, or looking at what the final idea is like and taking a wide and fuzzy view of the effect of one on the other.

To start with edge pieces, one of my favorite examples of the clash of music and science that addressed a very basic and important part of music was a classic psychological experiment carried out by C. F. Malmberg in 1918. He was interested in

---

★ Especially when playing Nigel Tufnel's "Lick My Love Pump."

trying to quantify and really nail down the idea of consonance and dissonance as a psychophysical phenomenon. Consonance and dissonance are psychological precepts underlying what combinations of notes in the Western twelve-tone system sound "good" or "smooth" versus which ones sound "tense" or "grating." At a core musical level, it's hard to get more basic than combining two tones, and yet the sonic and musical result is one of the most complex aspects of psychoacoustics that we have a relatively solid handle on. In Western intonation, the played musical scale is based on a mechanically linear separation of notes divided into twelve half steps until reaching the octave. The octave represents the same pitch, but doubled in frequency, and the musical range is composed of a series of octaves spanning from the infrasonic C_1 of a pipe organ (at 8 Hz) up to the piccolo's upper limit of about C_8 (around 4,400 Hz). However, the notes on this scale are not what you might think of as evenly spaced. Intervals are defined along a logarithmic scale, and the psychological qualities that emerge from combinations of sounds, such as whether they are major or minor, consonant or dissonant, emerge from the mathematical ratios of their base frequencies. Certain intervals have always been described as consonant—sounding smooth and lacking tension—such as unison (a note played with itself, yielding, of course, a base frequency ratio of 1:1), an octave (with two notes of the same pitch but one double the frequency of the other, yielding a 2:1 ratio), or a major fifth (with a ratio of 3:2). Intervals such as diminished seconds (two adjacent semitones played together) or even more musically common ones such as minor thirds (with a ratio of 65,536:59,049) are described as dissonant or tense.

Consonance and dissonance would seem like a relatively

simple musical feature to try to characterize scientifically. We all have experience with intervals (both musically and in other daily sonic experiences ranging from birdsong to our own voices) and generally have standardized reactions to musical intervals, at least within our own experience. The problem is that even this relatively simple musical feature is not a stable one. Historically, certain intervals were defined as consonant. Pythagoras defined consonant intervals as the ones with the smallest interval ratios, so most likely he was including only intervals from the pentatonic scale (unison, major third, major fourth, major fifth, and the octave). In the Renaissance, major and minor thirds and sixths were included as consonant, although minor thirds were supposed to be immediately resolved into a major chord afterward. In the nineteenth century, one of the first great psychoacousticians, Hermann von Helmholtz, declared that all intervals that share a harmonic were consonant, thus defining only seconds and sevenths (intervals adjacent to the octave notes) as dissonant. To add to the complexity, by the beginning of the twentieth century, musicians were experimenting with a much broader range of musical combinations and regularly using and sustaining intervals that would have made a Romantic-era composer pray for sudden-onset hearing loss.

This was the background for Malmberg's experiment. While the famous paper describes the basic aspect of getting musicians and psychologists to agree to a standardized scale of consonance and dissonance, the background story I've gotten from some older colleagues and Carl Seashore's classic 1938 book *Psychology of Music* was a bit more interesting (albeit possibly apocryphal). Malmberg deliberately chose a jury of musicians with no scientific training and psychologists with no musical training and forced them to listen to intervals played on tuning forks, a pipe organ, and a piano under strict environmental conditions (which one of my older sources insisted meant "no access to food or bathrooms during the session"). These sessions would be continued for a whole year or until *everyone* in the group agreed on a single judgment of consonant or dissonant for each interval.* And thus he acquired one of the first group psychological assessments of musical qualities in a study that is cited to this very day.

Despite the social complexity required to get a single critical chart that related the psychology of the perception to the underlying mathematics of sound, the study didn't dive too deeply into the underlying neural basis. It wasn't until we had a greater understanding of some of the biology of the cochlea that some particularly interesting relationships between musical intervals and neuroscience popped up. In 1965, R. Plomp and W. J. M. Levelt reexamined consonance and dissonance, using a similar psychophysical technique (i.e., playing simple intervals and asking subjects to rate them) but with a different context. Rather than just looking for agreement between musically trained and untrained groups about the psychological percepts of consonance and dissonance, they plotted relative consonance and dissonance against the width of the ear's critical bands.

Critical bands (the term was coined by Harvey Fletcher in the 1940s) are the psychophysical elements that almost made me ditch grad school and go back to being a dolphin trainer, largely because they were always explained by psychophysicists. Yet once you get past the jargon, they are pretty easy to understand.

---

* Just read two psychology papers on the same subject with different authors and you can get a feeling for how hard it is even to get *that* group to agree on anything.

If I played a 440 Hz tone for you, and you had some musical experience or perfect pitch, you could identify it as a single tone, a concert A in musical terms. If I played a 442 Hz tone, you could not tell the difference between that and the 440 Hz tone. Nor 452 Hz, nor 475 Hz. You probably couldn't tell the difference until I had actually changed the tone by about 88 Hz. This estimate is based on a filtering function of the hair cells laid out along the cochlea, with hair cells near the basal end (i.e., near the opening to the inner ear) responding to high-frequency sounds and those at the apical end responding to lower-frequency sounds. Although a healthy human cochlea has about 20,000 hair cells spread over about 33 mm (about 1¼ inches), hair cells within about 1 mm of each other tend to be maximally sensitive to the same general frequency. These regions that hear about the same frequencies are called critical bands. They are referred to as bands because they are roughly linear in the way they're distributed along the cochlea, but the critical bands have different frequency spans or bandwidths. Lower-frequency sounds, especially those in the vocal and musical ranges, tend to have much smaller critical bandwidths, yielding better frequency resolution, whereas pitches higher than 4 kHz or so have much broader critical bandwidths, meaning it is harder to resolve individual pitch changes across the same frequency span.

This gave Plomp and Levelt a biological basis against which to compare consonance and dissonance. Once again they asked their subjects to rate different intervals for consonance and dissonance for five different base frequencies, but they plotted these ratings against the width of the critical bands for these frequencies. What they found was that there were specific relationships between the rating and the position within the critical band. Very small differences in frequency between the two tones,

ones in which the two tones were separated by less than a full critical bandwidth, yielded the greatest dissonance rating, whereas those with a separation around 100 percent of the critical bandwidth separation were judged more consonant. In other words, listeners responded to different musical intervals based on the underlying organization of the hair cells in the cochlea.

The idea of consonance and dissonance has been traced upward throughout the brain. Research has shown that auditory neurons respond more strongly to consonant intervals than dissonant ones, and that the relationship between the two actually matches that seen in the older behavioral tests. The same has been shown to be true for higher regions of the brain (not merely auditory) that process both sound and emotion, with multiple studies showing differences in activation of emotional processing regions based on whether the interval is major or minor. This confirms what we know from listening on our own—that major intervals sound not only consonant but "happy," whereas minor ones sound "sad."

A recent study by Tom Fritz and colleagues took this one step further, in a paper titled "Universal Recognition of Three Basic Emotions in Music," in which the researchers examined the ability to recognize emotional aspects of music across cultures. They had Western (German) and non-Western (members of the Mafa ethnic group from northern Cameroon) listeners categorize music from the other's culture as happy, sad, or scary based on underlying consonance or dissonance. Both sets of listeners were supposed to be naive about music from the other's culture, to try to rule out familiarity and experience effects. The results showed that both groups identified the emotional content of the Western music, with similar directionality of "liking" the music in both the German and Mafa groups. It would seem as if we have a relatively solid, long-term, cross-cultural lock on the idea that there is a biological basis to music.

Or do we?

As with any attempt to address something this big with precision, there are a lot of problems. For example, in the Plomp and Levelt study, while they were using very precise intervals, the base frequencies that they used were based not on musical notes but rather on easily generated pure tones at specific frequencies, specifically 125, 250, 500, 1,000 and 2,000 Hz. Yet in musical contexts, you never run into pure tones. Every musical instrument creates a complex timbre (often called sound color) that includes harmonic frequencies. Even the simplest of flutes doesn't create pure sine waves, which are, in effect, what Plomp and Levelt were testing with. An excellent player on a world-class flute still creates not sine waves but triangle waves: in scientific terms, multiple individual sine waves add up to create odd-numbered harmonics that get reduced quickly at the high-frequency end, creating a pure-sounding tone with just a hint of bite to it. These harmonics are not simply ignorable overtones that arise from the structure of the instrument and the player's performance. The clashing and overlapping of harmonics are another important factor in the creation of perceived dissonance or consonance.

Next, the very idea of consonance not only has shifted across time within Western culture but has a great many problems when you start bringing in other cultures. For example, in the study using the Mafa listeners, the study couldn't present Mafa songs that were categorized as happy, sad, or scary because the Mafa do not assign specific emotional descriptions to their music. The German listeners had to decide on the emotional tone of the music based on the relative "dissonant" shifting of the key of the original piece. But if you actually listen to some of the traditional Mafa flute pieces used in the study, while the timbre of the flutes is somewhat coarse (and hence musically quite interesting), the actual intervals played are those found in Western twelve-tone scales. What would happen if you tried using music from non-Western cultures that employ different intonation and intervals? While all known human music includes the basic intervals that are commonly labeled as consonant, many go far beyond them. Indian ragas often use intervals of less than a Western semitone, and some Arabic music uses quarter-semitones. One of the more extreme examples (for most Western listeners) is Indonesian gamelan music, which uses either a tuning with five equally spaced notes per octave or one with seven unequally spaced notes per octave. Gamelan music sometimes not only uses both tuning techniques in a single song but also will slightly detune two instruments that are technically playing the same notes. This detuning creates a beating or roughness between harmonics, which is one of the definitions of dissonance for complex timbres.

Gamelan music is quite beautiful, but it takes some getting used to before you can appreciate it—the fine structure of the complex intervals and partial overlap of harmonics, particularly when played on percussion instruments, would make most naive listeners wonder who killed the piano tuner. Several years ago, I was co-teaching a class on musical perception and a student who was a devotee of gamelan questioned the very idea that the Western twelve-tone scale was really hardwired. To test this question, she gathered students who were not musicians themselves and had never heard gamelan music and had them carry out a traditional consonance/dissonance rating, comparing snippets of Western music and gamelan. Not surprisingly, the students showed the expected ratings for consonance and dissonance for the Western musical intervals and pieces but rated most of the gamelan music as dissonant or unpleasant.

The student then took another two groups but had them listen to a half hour of gamelan music before doing their ratings. She found that the students who had listened to the gamelan music still rated the Western music the same way, but significantly increased their ratings of the gamelan music as more consonant and pleasant. So with only brief exposure to an alternative tuning system, the supposedly hardwired system started rewiring, changing the psychological impact of what was heard. This led me to carry out an ad hoc experiment of my own without having to go too far afield musically. I typically ran the consonance/dissonance study as a demo in class, usually with twenty to thirty Brown undergrads, but I never separated out the musicians from the non-musicians. When I finally did, I found that those with musical training almost always shifted the scale towards consonant across the board, and among the jazz studies students, it was almost impossible to get any dissonance rating at all. Experience, at either a cultural or personal level, probably didn't change the distribution of firing from the hair cells in their ears, but it certainly changed the higher-level assessments of the intervals. None of this means there is no biological basis for the musical chords we recognize and the way we respond to them, but it certainly raises warning flags about making broad statements about music and the brain. It's important to remember that the thing being used as a measure, our own emotional responses, arises from a very complex infrastructure in the brain.

If the complex aspects of cognition are getting in the way of trying to find a "bottom-up" way to address music, another way to get a handle on the interaction between music and the brain is to try a more gestalt, "top-down" approach (to return to our jigsaw analogy: look at the picture on the box and try to match it) and see what kind of effect actual music, rather than a tiny acoustic subset such as a specific interval, has on the brain.

For example, almost everyone in the Western world is subjected to the seasonal terror that is Christmas music. If you listen to Christmas songs, especially popular-music ones, what you will hear is almost pure consonance. The intervals are almost always octaves, major thirds, major fifths, and major sixths with a few daring quickly resolved minor intervals, but all in all, these are songs to make you feel cheerful, relaxed, and happy. All of the songs have what would be termed "positive valence" by someone trying to calculate the statistics of an fMRI study while wearing a little red elf hat. Christmas carols emerged from the efforts of small groups of people singing a capella to celebrate the end of the harvest and the holidays. (A recent study showed that amateur singers tended to show a decrease in cortisol, a marker of stress, while professional singers showed the opposite; both groups showed an increase in oxytocin, a neurotransmitter implicated in arousal. This may be why you see highly enthusiastic off-key carolers plaguing the streets, while professional musicians often put out canny, cynical-sounding Christmas singles.)

Christmas songs have been around for more than six hundred years, and the basic musical tradition of consonance and simple tempos repeated at seasonal intervals has created years' worth of memory traces and emotional-auditory associations with holiday events. But there are caveats to even this seemingly innocent use of music to affect our emotions. If you hear "Santa Claus Is Coming to Town" when you're a kid, it becomes associated with the holidays, getting presents, and general family cheer. When you get a bit older, it can act as a pleasant reminder of happy times past. But after a while, you hear the first few notes and your brain says, "It's that song again." You tune it out. And shortly it becomes environmental noise, which is in fact more of a stressor than an emotional conditioner.

The phrase "familiarity breeds contempt" has a certain neural basis. As all of us who have almost reflexively rolled our eyes while being forced to listen to "The Twelve Days of Christmas" would agree, overpresentation of any stimulus often leads us to ignore it. While the *idea* of consonant music works in general at a population level, constant repetition begins triggering habituation.★ Habituation is characterized by diminished responses to the same stimulus over multiple presentations. This is why you don't usually startle at the second loud sound you hear in short order. It also kicks in faster and lasts longer if the stimulus is repeated more rapidly (with a shorter interstimulus interval, or ISI, in neurospeak). So after you survive the holidays and have nine months or so to recover from hearing "Frosty the Snowman," "Rudolph the Red-Nosed Reindeer," and other consonant offerings, when the season rolls around again, the first few times you hear these songs it works again, bringing on the holiday spirit, but as soon as the barrage of Christmas music gets going, so does habituation. And while every new singer and band hopes that *their* rendition of a Christmas song will be different enough to become a classic, it's pretty unlikely. Songs in a specific genre follow compositional rules, and when you're trying to keep a song Christmassy and recognizable as such, you have to

---

★ At the neural level, habituation is one of the simplest forms of non-associative learning and can even be observed in organisms without brains.

go pretty far afield acoustically to break away from the eventual habituation and stressor-based annoyance that most constantly repeated consonant songs impose.

But habituation to music is not just a seasonal malady. Anyone who has been stuck in a waiting room, elevator, shopping mall, or gas station has been subjected to music that would undeniably be described by psychologists, neuroscientists, and marketing people as having "positive valence." Often tagged with labels such as "easy listening," "light jazz," or the more blatant "elevator music," this type of music consists primarily of consonant intervals in major keys, with only occasional forays into the minor keys for flavor and texture, and is driven by regular, relatively slow tempos that you can tap your fingers to but which are probably unlikely to make you want to get up and dance (which would wobble the elevator). Presenting music with almost no auditory stressors in it is a widespread technique for reducing tension not only in individuals but in groups of people. First tried out in 1936 by the Wire Radio Corporation and eventually renamed Muzak, this type of music became so successful and widely distributed that the name has become a generic term. Despite being known for music that would deliberately not grab your attention by featuring anything with unusual keys, tempos, or intervals, Muzak was one of the first forays into consumer manipulation using music, and a wildly successful one. Purchased by Warner Bros. in 1937, long before the heyday of neuroscience, the Muzak company began formulating the first algorithms for changing group behavior by setting very specific limitations on tempo and tuning. It took this further by introducing what it called "stimulus progression"— piping in music with different tempos at different points in the workday and introducing structured periods of silence to limit habituation to the music. It wasn't until the late 1980s that they started broadening out into a wider variety of music, creating the first customized playlists for clients. But at that point they were up against the advent of portable customized players, digital recorders, and the beginnings of Internet music and radio, and Muzak Holdings filed for bankruptcy in 2009. Still, the legacy has lived on, particularly in places where acute stress can be a problem, and where you hopefully won't be long enough for habituation to kick in.

Think of the music you hear in customer service areas, emergency rooms, doctors' offices, and dentists' waiting areas—it is inevitably "easy listening" in some form or another. In an environment that is filled with anticipatory stressors, using highly consonant music may not make you feel better about whatever brought you there, but it is more likely to reduce your stress than a monitor showing the news. There have been hundreds of studies validating the ability of light music to reduce psychological stress. While most of these studies have been carried out using surveys in a relatively ad hoc manner, the findings have yielded some practical benefits. One study was carried out in a hospital emergency room in England faced with an interesting budgetary choice. Having limited funds, they were forced to choose between installing an easy-listening music system and hiring new security guards. What they found was that by electing for the music, they had significantly fewer hostile incidents involving waiting patients or worried families and did not have to resort to beefing up the number of security personnel. There have also been interesting clinical studies that showed that pregnant women waiting for amniocentesis (a common but stressful test that samples amniotic fluid to make sure the fetus is developing normally) who were exposed to music, as opposed to those who sat reading magazines or just sat, had significantly lower levels of serum cortisol, a metabolic indicator of stress.

But, as in any field trying to tie together complex neurally based behaviors and music, sometimes the research bandwagon picks up a lot of hitchhikers. A search on PubMed for "music" and "behavior" brings up almost 3,000 articles, many of which contradict each other and/or are clearly statistically challenged. One group of studies claimed that patients in a psych ward exposed to hard rock or rap acted out more than those exposed to easy listening or country (with no indication as to which songs, how loud, or why anyone would play highly arousing music of any sort to inpatients). Almost 1,900 papers can be found on the effect of listening habits on adolescents (most of which claim, without worrying overmuch whether their statistics actually prove it, that if your kids listen to heavy metal they will have low grades, display behavioral problems, be promiscuous, abuse drugs and alcohol, and of course get arrested).

These studies rarely add much except confusion to the question of how music affects human minds at a group or population level because they are treating music reductively, as a simple tool. It is very rare to find a study that even controls properly for basic elements such as loudness, repetition rate, or context effects. And the problem of working with groups is that you have to take into account individuals' history and experience, which, aside from basics such as age, gender, handedness, and hearing health, are usually beyond the scope of most protocols.* It may be statistically valid to say that consonant music in major keys makes most people feel relaxed or happy, but at an individual

---

* You would be shocked to see how many published studies fail to even check if the subjects have normal hearing. I used to be.

level, if a person had a happy childhood in a household that played a lot of Public Enemy or Trent Reznor very loudly, he or she would probably find the valence of easy listening not so positive. So statistical findings based on generic people get reported in papers, and once the findings get out into the popular arena, things get even weirder. One of the best examples of this is the popular uprising of what started out as a reasonably interesting study on music and spatial reasoning and exploded through the press as a way to increase intelligence.

The "Mozart effect" emerged on the popular scene in 1993, but it had its roots much earlier, in the world of a French otolaryngologist named Alfred Tomatis. Tomatis proposed that auditory deficiencies too subtle to be defined as clinical hearing loss were responsible for a very wide variety of psychological and neurological disorders. His basic tenet was that early developmental problems in the ear led to broad-scale deficiencies in neurological processing of patterned input, such as speech and music. Inventing a system called the "electronic ear," Tomatis used a combination of filters and amplifiers to try to restructure sounds into areas of his patients' hearing range that he thought were deficient. His initial patients were opera singers, and his theory was that the sheer volume and stress of the extreme vocalization behavior in singing opera had damaged their middle ear muscles, thus reducing the protection those muscles offered against overloud and usually self-generated sounds.

There is a real basis to Tomatis's hypothesis: mammals (and some other vertebrates) have a reflex arc that clamps down on the tympanum and ossicles to dampen sounds that may be loud enough to damage hearing. However, the system is limited, in that the amount of time it takes for the reflex to kick in is too long to prevent damage from very fast-onset loud sounds such as explosions. Plus, like most reflexes, it habituates with chronic exposure, preventing it from being able to prevent hearing loss due to chronic exposure to loud noises. Tomatis's idea was that "the voice cannot reproduce what the ear cannot hear," and so his early electronic system tried to present sounds in the range that the patients' ears were no longer responding to optimally. Claiming success with these techniques, Tomatis expanded his work into other clinical areas, ranging from depression to autism. He had his patients listen to music with very structured tempos and specific registers, including Mozart symphonies and Gregorian chants, two types of music with very different pitch and tempo characteristics, but both sharing a great deal of harmonic content and temporal structure. He claimed numerous successes in treating these psychological and psychiatric conditions, following them up with a book called *Why Mozart?* in which he tried to explain why Mozart in particular was particularly useful for treating conditions of the mind by retraining the ear to hear properly.

While Tomatis was a highly prolific communicator about the power of music to treat a wide variety of conditions (writing fourteen books and several thousand papers), his work suffered badly from a lack of statistical rigor. Most of his papers were clinical case studies with very limited numbers of patients, focusing on individual successes rather than trying to establish basic principles of how the treatments worked (or failed to). In fact, his failure to carry out properly counterbalanced studies and provide detailed data was the major reason he began to break away from the medical establishment.

Most attempts to replicate his results using standard scientific techniques have shown only minimal or no effect. This is in part because of the march of scientific progress. His theories about retraining hair cell function in the cochlea have fallen by the wayside because more recent work has demonstrated that mammals, including humans, cannot regenerate hair cells. Once a certain band of hearing is lost, it stays lost; however, there have been some studies indicating that the brain recruits other frequency bands into the region that has the damage. This is why some people who have lost significant high-frequency hearing, a normal process of aging, are often unaware of the fact and are convinced that people around them are mumbling. Or, as my colleague Lance Massey (a veteran of too many nights with his ears stuck between high-powered studio monitors) said more succinctly, "I thought I could hear high-frequency tones until someone actually played some for me." In addition, Tomatis claimed his audio-psycho-phonology (APP) could treat a remarkable breadth and range of conditions including schizophrenia, depression, dyslexia, attention deficit disorder, and autism. His theory that all of these were based on failures of hearing has not been borne out by subsequent research. While such failures may underlie some psychiatric conditions, loss of hearing is not always or even usually the underlying cause. And yet there are still thousands of practitioners who utilize APP and its follow-up technique of sensory integration therapy with mixed results at a statistical level, based on a number of proponents who claim tremendous success.

You would think that with the failure of Tomatis's treatments to show success in follow-up studies and the normal progression of scientific theories, the idea of Mozart's music being somehow special would fade away. But then, in 1993, Frances Rauscher, Gordon Shaw, and Katherine Ky published a study in *Nature*, one of the world's leading scientific journals, titled "Music and Spatial Task Performance." The study examined the relative effect of exposure to ten minutes of Mozart's Sonata for Two Pianos in D major, K448, versus a relaxation tape or silence on thirty-six college students' performance on the abstract and spatial reasoning part of the Stanford-Binet Intelligence Scales test. The results of the study showed a relative increase of eight to nine points after listening to the Mozart music as compared to the other two exposures. The paper reflected results from a relatively minor study, but it did make the point that the effect was temporary and only reflected exposure to a single piece from a single composer and made no adjustments for subjects with musical versus non-musical training. Its primary claim was that there was a difference in spatial listening scoring on a standardized test after exposure to a specific type of highly structured, relatively complex music.

If this paper had appeared in some other, lower-impact journal, it most likely would have entered the literature to be occasionally cited in the context of spatial reasoning and music perception, an area that has quite a following and has made some interesting contributions to the notion of the interaction of music and mind. However, since it appeared in *Nature*, the trouble started almost immediately. In the next volume, a response to the study immediately pointed out problems with the statistics, throwing the study's results into question. Over the next few years, studies came out both supporting and reviling the original piece, some showing that the effect disappeared if the experimenters used different intelligence tests, others highlighting the specialness of Mozart in that exposure to Beethoven did not show a similar increase in spatial reasoning. Some studies even claimed that rats would show the Mozart effect, showing better maze navigation after being exposed to his music.

And here's where the conflict between science and popular

culture starts causing problems. The basic idea of the study, that "listening to Mozart makes you smart," was picked up by the popular press, with the *New York Times* publishing an article claiming that since listening to Mozart made you smarter, this made Mozart the world's greatest composer, and an article in the *Boston Globe* citing a study showing that giving your children classical music lessons made them perform better on intelligence tests (a study I have not been able to find). Articles spread throughout the emerging Internet and blogosphere, getting further and further from the original modest findings and thrusting the Mozart effect into popular culture with the same fervor that greets a new celebrity diet. The high point of the cultural adoption seemed to have occurred in 1998, with the then governor of Georgia, Zell Miller, introducing a line item in the state budget to give all children born in his state a recording of classical music to make them smarter, followed by the state of Florida passing a law requiring that state-funded day care centers play classical music, and an article in the *Houston Chronicle* reporting a mandate that Mozart be played to inmates.★

The Mozart effect has become a multimillion-dollar business. A quick examination of Amazon.com shows that there are almost 250 recordings (mostly for babies) based on the Mozart effect and promising to increase intelligence, heighten performance, improve concentration, and "heal the mind," along with about 900 books on the subject. Yet almost all subsequent research has shown that there is no such effect and that you can get the same non-significant result by doing anything you

---

★ I couldn't find any reports of an increase in prison escapes in Texas following this, which sort of suggests the Mozart effect is not significant in planning the spatial aspects of jailbreaks.

enjoy, including having ten minutes of silence (one of the controls from the first test) or listening to an excerpt from a Stephen King novel. This kind of conflict is the type of thing that makes editors' and music distributors' careers. Its journey from study to marketing device was even written about in a paper in the *British Journal of Social Psychology* titled "The Mozart Effect: Tracking the Evolution of a Scientific Legend."

But it is not the kind of thing that makes scientists happy, no matter what their field. Finally the original lead author had to issue a statement that they'd made no such claim that listening to Mozart enhanced intelligence and that the effect was highly limited to certain spatiotemporal tasks, finally commenting in an article in the *New York Times* in 1999 that the money Georgia's governor proposed spending on recordings for babies could probably be more wisely spent on music education. But as I know from personal experience, as soon as you start using terms like "spatiotemporal" in an explanation, you've lost 90 percent of your audience, who will go back to buying CDs with pictures of smiling babies and the words "smarter," "happier," and "creative" on the cover.

But is there any underlying basis to the idea that exposure to music can have a positive effect on task performance? Because of the way science works, most studies have continued to use the particular Mozart piece from the original study. There have been some interesting studies showing that while there is no real measurable effect on intelligence, listening to this particular sonata and Piano Concerto No. 23 seemed to reduce seizure-like activity in the brains of epileptics (although there is no indication that it had any beneficial effects in terms of reducing the severity of actual symptoms).

There have been several studies showing that listening to

Mozart's music in the background while carrying out visual tasks produces increased neural synchronization in what is termed the gamma band. The gamma band is a collective neuronal response observed in EEG studies from 25 to 100 Hz (usually around 40 Hz) that is emphasized not only during music perception but also during attentional processes and which has been implicated in what is called "the binding problem," the poorly understood underlying basis for integration of all brain activity into consciousness. There are literally dozens of sites in the brain that are responsive to music (Mozartian and otherwise) or even individual elements of musical stimuli and are also implicated in task-related behaviors. At a basic level, there is a greater tendency for pitch and musical processing to occur in the right hemisphere (of right-handed people, anyway), the cortical hemisphere that is more commonly involved in spatial and emotional processing. Specifically, while ventral (lower) projections from the primary auditory cortex seem to be involved in the identification of specific features of sounds such as pitch and absolute duration, dorsal (upper) projections from this area carry information about changes in frequency over time and may have links to motor systems, making them critical for not only perception but also tasks that require precisely timed motor behavior, such as the performance of music, dance, or even general locomotion. This may underlie some of the observed efficacy that music therapy seems to have on motor-related disorders such as Parkinson's disease. And while there have been clinical studies demonstrating that it is possible to separate out pitch perception from rhythm perception, neural imaging studies that have examined the auditory region of the temporal lobe of the brain have not consistently identified any specific location underlying identification of musical tempo.

What has emerged in many studies is that areas involved in the perception of musical timing are also deeply tied to motor behavior, including the cerebellum, which plays a role in fine motion coordination; the basal ganglia, which are important in both voluntary and procedural learning behaviors; and the supplementary motor area, which is involved in the planning of motor behavior.

None of these findings means that Mozart knew some neuroscientific secret that he incorporated into his music. The predominance of Mozart's music in many studies is based on two things. First, scientists tend to cluster around stimuli that have shown an effect in previous studies (it also lets them cite previous work in their own). Second, Mozart's music (as well as that of Bach and other late Baroque through early classical composers) tends to be based on relatively simple repeated double phrases or three-part forms with simple interval structure (compared to contemporary music) and non-overlapping tempo (what's called "long-term periodicity"). In addition, the music was written to be performed on analog instruments—in other words, it was to be played at tempos manageable by human performers. So the music of Mozart (and other composers of similar style) may activate broadly overlapping neural structures that help with spatial, attentional, motor, and other processes.

In short, it's not the specific composer, genre, key, consonance, or rhythm of the music that may have an effect on a person; it's the fact that embedded in the music, whether heard or played, are the underlying rhythms and processes that drive the brain in day-to-day behavior. As Robert Zatorre, one of the leading lights in neuroscience research in music and the brain has pointed out, "The continued interactions of musicians and scientists will be important, as the study of music and neuroscience is *mutually* revealing." As we bring more and more technologies and ideas to bear on the twin processes of understanding music and understanding the mind, highlighting the biological aspects we can explain and hypothesizing about those we can't, we can approach a sort of "tipping point" for acoustic awareness—how your mind is shaped, and sometimes manipulated, by sound.

Chapter 7

STICKY EARS:
SOUNDTRACKS, LAUGH TRACKS,
AND JINGLES ALL THE WAY

O NE OF THE first things you have to think about when you
start a career in science is whether your focus is going to
be on basic or applied research. People who want to do basic
research are those who like puzzles—they pick a specific problem
and want to solve it. Those who are interested in applied research tend to want to see their work help solve some real-world
problem. But the wonderful thing about such a universal field as
sound perception is that even the most obscure aspect can end
up having real-world applications. It's just a question of what
part of the world you're improving, whether it's using the psychophysics of audiograms and critical bands to develop the
MP3 compression algorithm or doing detailed recordings of the
environment around someone's head to create surround sound
systems. But our most powerful research tools for auditory science are not fMRIs or EEGs—they are our ears and our brains.
And there is a very long history of applying auditory principles
to things in our everyday life to get directed emotional or attentional responses from listeners.

One of the most common large-scale formats for controlling

emotional responses is movies. Aside from the kind of blasé educational films we are forced to watch in school, almost all films, TV, video games, and other multimedia are about directing attention and inducing an emotional response in the audience. While the first films were developed around 1880 by Eadweard Muybridge, an experimental photographer known mostly for his pioneering work in analyses of animal and human locomotion, the first films shown to a general audience were presented in 1893. They were totally silent. If you have access to a film library showing some of these very early silent films, watch them and try to figure out why they come across as kind of flat.* The complete lack of sound removes some essential underlying drive, which is why theaters started adding live musical accompaniments almost immediately, ranging from live piano players in small houses to pipe organs in the larger ones. Once movies became a serious industry, original scores were often created to be performed by entire orchestras as a real-time soundtrack for the film, starting with Joseph Carl Breil's score for D. W. Griffith's *Birth of a Nation* in 1915 and, arguably, culminating in Gottfried Huppertz's score played for the premiere of Fritz Lang's *Metropolis* in 1926.†

It might seem odd to think of movies as applied neuroscience and psychology, but "neurocinema" is very trendy these days as a tool for examining how people's brains react to films in order to improve their effectiveness. One approach that's be-

---

* Check out the Edison motion pictures collection from 1891 to 1898 at the Internet archive: www.archive.org/details/EdisonMotionPicturesCollectionPartOne1891-1898.

† There have been ten different scores played with this film, each providing a different emotional basis and take on the story, and often quite specific to the time in which each was created.

ing taken by a San Diego group called MindSign Neuromarketing is to have people watch a short section of film and then examine their brain blood flow in an fMRI, usually focusing on the amygdala after watching a horror film. This is a very sexy approach, but it has some important limitations. For one, it's very expensive. For another, in order for researchers to gather data at many points during the movie, the subject has to watch very short chunks, which interrupts the flow of the film and therefore limits the value of the information. In addition, a single fMRI scan takes at least two seconds, during which you go through several hundred perceptual events. So the data are neurally fuzzy and unlikely to yield anything more accurate than what you'd get by watching a filmgoer's face and movements during a movie.

The soundtrack in film or video is composed of both the film score (the underlying music that provides an emotional timeline for the work) and the dialogue and environmental sounds that drive the narrative and environmental immersion in the film. Both score and full soundtrack can provide a lot of power in terms of driving your attention, controlling your arousal, and, filmmakers hope, making you remember the film and the emotions you felt while watching it. But the score and the soundtrack have different ways of accomplishing these goals.

The score of a good film uses music at what it does best— providing an underlying psychological flow to the story without having to give specific information about the action or the environment. One of the most basic techniques is to use a theme song or anthem to provide identification or "branding" of the film or its elements. An anthem uses properly composed music to ensure strong emotional association with the visual and narrative aspects of the film. Think, for example, of the low drums and forbidding horns of the "Imperial Death March," which accompanied Darth Vader wherever he walked. When you heard that music, you knew there was trouble.

Music is one of the strongest associational tools available that doesn't use a bunch of levers, lights, and electrical shock equipment. What is often the first thing you remember about a show you saw when you were young, even if you can't quite put your finger on the name? It's usually the theme song. A successful musical theme grabs your attention and locks you into the whole experience of the film or TV show in a short time, usually under seven notes to make best use of your limited short term memory span. For example, if you were born after 1955 and before 1975 and I even mention the 1960s television series *Batman*, I bet the back of your brain starts singing "na-na na-na na-na na-na na-na na-na na-na na-na na-na Batman!" in the appropriate key. And unless you are a serious classical music listener or musician, when you hear five slow notes accompanied by images of apes worshipping a strange black obelisk, you think *2001: A Space Odyssey* rather than Strauss's *Thus Spake Zarathustra*. Even more minimally, think of the two alternating notes that made up the basic theme to the classic spaghetti Western *The Good, the Bad, and the Ugly*. The brilliance of Ennio Morricone's score for this film was that not only do you think of that film when you hear this three-second clip, but by using different instruments (a flute, an ocarina, and human voices) consistently before showing different characters, you form an audio and emotional association with the characters even within the context of the story just based on the timbre, the fine structure of the tones, separate from the actual pitch. Perhaps the ultimate in the use of anthemic music was John Williams's score for the *Star Wars* series. He used no fewer than eight separate anthems just in the initial film (*Episode IV—A New Hope*) and twenty-six separate ones across all six films.

In all of these, the basic strength of the theme or anthem is based on the limited number of sounds, the relative simplicity of the interval arrangement, the consistency of presentation, and the proper use of repetition (e.g., as an acoustic bracket at the beginning and end of *2001*; when the appropriate characters appeared on camera in *The Good, the Bad, and the Ugly* and *Star Wars*; and every week, same bat time, same bat channel, for Nelson Riddle's *Batman* theme).

Another effective and well-known anthem in a movie was used in 1975 in the film *Jaws*. I'm betting that as soon as you saw the title, you started hearing that bass ostinato, a low-pitched heartbeat-like sound that grew in loudness as it repeated; deep down in your brain stem, you *knew* something was going to happen that would not involve butterflies and unicorns. The *Jaws* theme, written by John Williams, is one of the great iconic film score themes, and it's played on, of all things, a tuba. I get a little cognitive dissonance when I think of that; how could a tuba be threatening unless it was thrown at you? I had a friend who played the tuba professionally, and while she was very good, watching her turn purple as she blew into the mouthpiece while wrapped up in eighteen feet of brass and valves was about as threatening as someone having an asthma attack in a radiator (she eventually switched to electric guitar and rocked out). But the tuba has several things going for it as an emotional driver, at least when it's not relegated to the oompah section of a march-ing band. The tuba is a very low-pitched instrument, with the bottom end of some models going down into the infrasonic region.* Perceptually, lower-pitched means larger. This is based on very basic biomechanical and evolutionary principles that work in almost all vertebrates. A larger animal will have a larger vocal apparatus and larger lungs to make louder and lower-pitched sounds. In the world of females' mate selection (like the bullfrogs we spoke of earlier), a larger mate likely will have healthier genes to pass on, or better survival skills at getting resources if it is a social animal.† But outside the context of mate selection, an animal whose vocalizations are louder and deeper in pitch than others' is something that's bigger than you and is not worried about trying to conceal its presence. In fact, one theory of why large animals roar is that the sound deliberately induces a startle in their prey, to get them to move and hence start the chase, instead of the larger animal having to sit and wait them out.

So using a tuba for the opening theme of *Jaws* was a perfect example of how sound should be used in film, drawing on basic psychophysical principles as well as higher-order associations. The basic structure is a slow heartbeat-like pattern, speeding up until it matches a racing heart. Being exposed to an auditory signal that resembles important biological patterns can induce what is called "auditory facilitation." An example is an experiment I ran a few years back in a small gallery-like space, in which I played pink noise pulses (with long ramp-on and ramp-off times to make them sound like breathing) very quietly through hidden speakers and monitored the breathing rates of everyone who en-

---

* My tuba-playing friend insisted that tubas don't play major and minor scales, they play Richter scales.
† In the lab it was called the "Barry White effect."

tered the room.* Consistently 80 percent of the people exposed to this simple environmental change matched their breathing rates with the sounds they heard. It follows that if you play a respiratory or heartbeat-type pattern, first slowly and then faster and faster, listeners' breathing or heart rate too will start racing. So thirty seconds of slowly speeding-up tuba was a perfect way to get an audience to the edge of their seats, wondering what was really going on, when all they were looking at was a woman floating in the water. In short, a good anthem is one that obeys the psychophysical rules for auditory association and learning.

But unless you have a very rigid schedule and spend a lot of time composing specific playlists, real life rarely has a musical soundtrack. What you have out in the world is an environment where the sound provides a background to your daily life, giving you reverberant spatial cues about the size of the space, event-based sounds such as rising winds to tell you that the weather is about to change, and traffic sounds to give you warning about areas to avoid in order to make it to your next destination. It rarely includes someone popping up from behind a hill to play a sad trombone sound when you drop your ice cream cone. While hundreds of millions of dollars are spent on trying to make films more immersive, using everything from surround sound and bass cannons to 3-D visual technology, seeing a movie or watching a TV program is still a limited sensory experience.† The director is trying to create an entire world using

---

* Pink noise is a form of noise where the amount of energy is equal in each octave, yielding less noise as you go up in frequency. It has a more natural biological sound than white noise, which has a flat power spectrum across all frequencies.

† I deliberately omit John Waters's "Smell-O-Vision." I suggest you do too.

only two of your five senses. There are numerous tricks that can be used, such as distorting the motion at the edges of the screen, which causes your peripheral vision to trick your vestibular system into thinking that you're inside a swooping fighter jet. But you still can't smell the dust of a desert, taste the martini the lead character is sipping, or touch the fur of the dog on the screen. Until we build holodecks, a film is limited to your telesensory systems, and so the use of well-composed, artfully designed, and skillfully mixed background sound is critical in enabling the suspension of disbelief by enhancing viewers' spatial and emotional responses.

The heart of this is the use of sound effects. Most people who think about sound effects think about the work of Jack Foley, who in 1939 developed mechanical and analog sound effects to be synchronized with on-screen action. But the first major definitive work on sound effects was printed several years before Foley came on the scene, in the BBC's *Radio Times* yearbook for 1931. This article explained how to properly use sound effects in radio dramas, and despite being written eighty years ago, it defined the different categories of sound effects in a remarkably scientific manner, spanning everything from realistic event-related sounds to emotionally evocative ones. The parameters were widely adopted in various forms in the United Kingdom and United States and are still used to this day.

But these guidelines didn't emerge solely from the creative minds behind the radio shows. There was a major wave of scientific discovery in psychoacoustics in the period immediately before the widespread adoption of radio theater, the most commercially important form of media available in homes before the advent of television in the 1950s. The 1920 and 1930s were a period of major upheaval in the science and engineering of sound, providing not only the infrastructure that allowed the relaying of radio into private homes but also enough of an understanding of the psychophysical basis of sound perception to let people imagine entire story-based environments even through a small, frequency-limited mono speaker. During this period Georg von Békésy published major works on the propagation of sound in space and how sound gets distorted in both the environment and the ear. Harvey Fletcher, the father of stereo recording, carried out many of his basic experiments in speech intelligibility, examining the perceptual basis of loudness and noise rejection and leading to the theory of critical bands. Three Bell Laboratory scientists—Vern Knudsen, Floyd Watson, and Wallace Waterfall—gathered forty other physicists and psychoacousticians and formed the Acoustical Society of America. The 1920s through the 1940s were the heyday of understanding how we perceive the acoustic world, and its most obvious contributions were arguably not the wealth of scientific papers but rather the ability to actually use sound to create worlds through popular media.

There are some moments in film that have stuck with me—something grabbed my attention, made me feel a certain way, or in some cases made me jump out of my seat. I usually find after a second viewing that it's because something very cool or odd happened with the sound.* A properly designed soundtrack,

---

* My tendency to stop a video after a particularly cool sound effect, record the effect into a sound capture program, and subject it to spectral, temporal, and phase analyses is one of the reasons my friends don't like watching videos with me. Except those who do the same thing.

combining the film score, dialogue, and sound effects, has to weave together the psychophysics of spatial perception; ensure proper temporal alignment of visual and auditory events to get multisensory convergence; provide dialogue to propel the narrative along; and use music and evocative sounds to move the audience's emotions. The cost of failure is an almost immediate psychological glitch in the viewers; for example, the sudden blare of a musical theme grabs your attention and pulls you out of the story, leaving you saying, "What was that?" A friend of mine who does film composition stated it clearly: "The best soundtracks are the ones you don't even notice are there." It all depends on what you want your audiences to feel.

An easy approach is to go for an emotionally primitive response in the audience. Think of virtually any of the current crop of action films, ones that seem to spend a good chunk of the film showing explosions, car chases, gunshots, and robots turning into trucks. The soundtracks for these films consist of very loud sudden-onset sounds and rapid-tempo, loud musical scores with a lot of high-powered bass lines and dissonant intervals, with little or no break; in short, the sound is designed to get you aroused and keep you there with trying to split your time between startles, fear, and excitement. At a psychological level, this is the low-hanging fruit of auditory responsiveness. It's all about sending traffic back and forth between the brain stem and elements of the fear and arousal system, getting your sympathetic nervous system geared up for fight, flight, or sex (the last of these being the polite term for the third "f" of sympathetic activation). But the problem is that the only way to extend this response over the two hours of a film is to continue making the soundtrack louder and more dissonant, leaving you at the end of the movie all wired up, but missing quite a few hair cells.*

At the other end is the use of silence. One of the films that always struck me as incredibly powerful was Stanley Kubrick's *2001: A Space Odyssey*. It seemed to me that every use of sound in that movie was iconic, from the use of classical music to highlight motion and movement to the ambient, almost tempoless flowing frequencies of Ligeti's *Requiem* during times of tension and contemplation. But what I found most interesting was how the film used the *lack* of sound. Not just the absence of dialogue at the beginning and end of the movie, not just the lack of overlap between music and dialogue at any point, but rather the proper use of silence in the space scenes. When the deranged computer, HAL, activates and sends the pod under remote control to attack Frank Poole, followed by the scene of him hurtling out into space, there is nothing. No sound. No music.

As a space baby, I was raised conflicted.† I knew from all my early astronaut-wannabe childhood years that there was no sound in space, yet Saturday afternoons were filled with science fiction movies of the 1960s and 1970s, which always had spaceships going *whoosh* and hearing the explosions of whatever space weapons they were using to kill the Thing from Venus (although I do confess to having a soft spot in my heart for the elephant trumpet/car screech that the *Star Wars* TIE fighter made when it screeched by). Perhaps the most well-known example of this

---

* There have been several studies carried out on current films that have shown that the loudness in most major theaters is well above suggested levels for being able to hear the next movie you see.
† I was born eleven months before Yuri Gagarin's first space flight, and I hate the term "baby boomer."

conflict was described in an interview with Gene Rodden-
berry, the creator of *Star Trek*. Being technically minded,
Rodenberry knew that the *Enterprise* wouldn't make a sound as
it swooped past the viewers in the opening sequence, but he had
the sound team add in a swooshing sound because he thought
that otherwise it felt too flat. At a psychophysical level, he was
right—we've evolved to expect dynamic events to be associated
with sounds. A large object moving by us at high speed (without feet to make footsteps or wheels to make road noise) would
move a large body of air out of its way, generating a noise-like
band, shifting its center frequency up as it approached and
down as it moved away because of the Doppler effect. A lack of
sound would make us feel strange, kicking in the anticipatory
heightening of auditory and attentional sensitivity that silence
triggers even at a neural level. That is what, to me, made that
moment in *2001* so unique—by using the lack of sound an ob-
server would experience if he or she was actually in space,
Kubrick took us viewers out of our normal environmental context and put us in a soundless void, bringing along all the tension and attention that silence carries.

But while big-budget movies can usually count on being
shown in a theater with a decent sound system, what if you are
relying on whatever system and environment people have at
home? Sound design for the small screen is different. In a movie,
if you want to add tension, you can throw in atmospheric music
or the sound of engines with a serious 40 Hz rumble in it and
shake the theater, triggering the listeners' fight-or-flight system. But unless the person has a really good home theater setup
(which wasn't even available at the affordable consumer level
until the 1990s), those audio effects are lost. Sound design for
the small screen takes a bit of widgeting, as it has to use the

middle-of-the-road technology and middle-of-the-range sound quality that most common home systems have.

Luckily, even inexpensive systems these days are stereo and have a sound quality that is far beyond just adequate in the best human hearing range, so by using some psychophysical tricks, you can get some very powerful effects even through a home TV set. For an example, one of my favorites occurred in an episode of the 1990s show *The X-Files*. The show was highly successful, spanning nine seasons and spawning two movies. I liked it, but not for its acting or originality—many of the episodes were serious homages to previous mystery/conspiracy shows such as *The Twilight Zone* of the 1960s or *Kolchak: The Night Stalker* of the 1970s. I (and a lot of people I've spoken with) liked *The X-Files* because its atmosphere was genuinely creepy and emotionally powerful, largely due to the efforts of composer Mark Snow, sound editor Thierry J. Couturier, and sound designer David J. West.

The show used many typical auditory tricks—low-pitched strings, sudden silence, characters speaking in noisy environments—but one that particularly stood out for me was a specific scene in which the character Fox Mulder was speaking to his partner, who was supposed to be a maternity patient in a lab, and for some reason it was incredibly tense. I kept watching the scene, repeating it, switching to headphones, finally breaking down and recording the background sounds and removing the dialogue, trying to figure out why this rather banal scene seemed so incredibly menacing. When I finally examined a continuous piece of the background sound, I found that what I thought was supposed to be air-conditioning noise was in fact the expected noise convolved or mixed with the sound of a nestful of angry wasps. A wasps' nest is one of those sounds that doesn't require interpretation; it gets right into your brain stem and lets you know at a very basal level that this is not a good place to be if your desire is to remain in a pain-free state. By sampling the sounds of wasps, a fear-inducing primitive sound, and pushing it way down in the background but still leaving it perceptible, the sound designers were able to manipulate the scene to create a sense of tension and fear far beyond that elicited by the visuals or the situation.

Another aspect of audience manipulation that is specific to radio and the small screen is the laugh track. It's a pretty simple thing—the "studio audience" (often a pre-recorded tape) laughs at moments that are supposed to be funny, and you are supposed to laugh along. The laugh track was supposedly born in 1948 on the *Philco Radio Time* show. A show was recorded when a comedian was having a particularly good night, but because of some rather off-color humor, the jokes couldn't actually be used on air. However, they saved the laughter from the audience on tape and reused it on other shows. On television, a CBS sound engineer in the 1950s named Charles Douglass started using pre-recorded laughter to tweak the laughter from a live audience, mixing in the taped laughs if the audience wasn't laughing hard enough, or muting everything down if their laughter went on too long. By the time the 1960s rolled around and most shows were pre-recorded without a live audience, canned laughter became a rather annoying default, not dying out until the 1990s; since then it has become a bit of a rarity.

This powerful, if often counterproductive, sound production tool relied on psychology. A laugh track relies on auditory social facilitation: the laughter is a social signal, often given off as a way of relieving stress (most basic comedic situations involve something unpleasant happening to someone else). Studies have shown that there are four basic states that are communicated in the acoustically complex signal of laughter—arousal, dominance, the sender's emotional state, and what the sender thinks the listener should be feeling (called "receiver-directed valence"). Certain aspects of laughter are similar to the prosodic emotional cues found in speech, making it a powerful non-language-based communication channel.

Perception of laughter is processed throughout much of the auditory and limbic centers, with different types of laughter processed in different regions. For example, the kind of laughter that is evoked by tickling is processed in the right superior temporal gyrus (STG), a region that is often implicated in social "play." Laughter from an emotional response is processed more in the anterior rostral medial frontal cortex (arMFC), a region implicated in emotional and social signaling. But neural imaging studies have shown that both the perception of laughter and the act of laughing itself can cause the ventromedial prefrontal cortex to release endorphins, the brain's native and powerful pain reducers; in fact, many studies have shown that laughter can actually elevate your pain threshold. Another interesting possibility is that synchronized release of endorphins may play a role in social bonding. So while the laugh track emerged from one comedian's particularly good night telling off-color jokes, its ability to induce social bonding through sound turned it into an industry standard for manipulating the emotional response of audiences.

But there is another form of media that uses sound to immerse and manipulate its audience, and only recently has it gotten the kind of attention paid to radio, TV, and movies. Many examples of this form use extremely short, simple sounds that produce rapid emotional responses among almost all listeners, and yet they remain unstudied and unused, probably because they're video games. Go play a simple video game, particularly those from the dawn of the home computer game era, when sound was limited to 8-bit boops and beeps and very simplified tonal phrases—something such as *Pac-Man*. A trio of chords going up means victory. A descending arpeggio means you died and should be sad about it. A noisy blatting sound (such as in *Q-bert*) means frustration. These sounds had to be as simplified as possible given the constraints of the technology of the time, yet still rapidly emotionally evocative enough to keep you involved in the game.*

But the tremendous strides in computing power and the lowering costs of components have led to video game sound design being a true leader in the quest for sensory immersion.† Today's video games are even more powerful bastions of emotional manipulation, using more subtle techniques that match the more powerful underlying technology and sound programming. I remember that in the late 1990s when I was stuck playing *Quake 2*, the sound was what really got me—the sound of flies around a dead body, voices supposedly of tortured soldiers with flat affect, repeating over and over. But the one simple sound effect that caused me the most trouble was a simple tapping sound made by the foot of one of the Strogg (bad alien guy) characters—a simple, quiet *taptaptaptaptap* repeated over and over quietly as you approached its position, the only thing that

---

* Please don't e-mail me about how the Commodore 64 SID chip rocked. I don't need to be reminded how old I am.

† Although, like many other gamers, I tend to turn off the music track—there seems to be less attention paid to using composition to carry game flow than there is to making the simulated environment realistic.

warned you that it was around. After spending about eight night-time hours thoroughly immersed in the game when I should have been working on my dissertation, I had to drive to work the next day. I was going along in my car when I suddenly heard the same *taptaptaptaptap* and found myself slamming on my brakes, just narrowly avoiding ramming into the eighteen-wheel diesel whose exhaust flap was tapping in the exact same pattern as the monster in the video game. I'm still not sure if the adrenaline rush was from not dying in a car crash or from realizing I wasn't going to be fragged by a Strogg in the middle of Peacedale, Rhode Island.*

All of these types of media—radio, movies, TV, video games—are based on the idea of using sound to create a world, to let people suspend their disbelief by using their ears to create an environment beyond the edges of the screen; the creators hope the experience will be attractive enough for you to want to pay for it. But sound is also used to create an emotional micro-world, one that you take with you—and one that the creators also hope will induce you to pay for the experience. The micro-world is called the jingle.

My friend Lance was anointed the most annoying person in the world by *Maxim* magazine. If you know him, this would be surprising—Lance is very kind, hardworking, and funny, and he spends most of his free time coming up with new and un-usual ways to get your attention with sound. But the numbers don't lie. You see, Lance is the creator of the T-Mobile ring-tone. Five notes that come as a default sound for more than 30 million people in the United States alone and which signal dinners interrupted, work overdue, or one more annoying but

---

* Don't even get me started about the turrets in *Portal*.

unavoidable phone call. Five notes that somehow never really leave your head. Five notes that you hear anywhere you walk, even now, when you can download anything as a ringtone. *Maxim* called it "an aural scourge on humanity." Lance says, "It's just a jingle."★

There may not be that much difference between the two. And not all of it is Lance's fault. Some of it's your brain's.

As a neuroscientist, I've always been fascinated by jingles in the same way I was fascinated by leeches. Jingles are sort of a microcosm of applied music in the same way that the nervous system of leeches is a functional miniature of a human brain. Jingles are just about the simplest form of music out there that have all the same hallmarks as whole songs, concertos, et cetera— they are composed of pitches at specific tempos, they manage to worm their way into your memory, and they evoke an emotional response. They just have to do all that in a few seconds.

You know what a jingle is—it's a horribly catchy piece of music tied in with a product or name, a bit of sound that you can't get out of your head for days on end and which you can find popping up years after the product being advertised is long gone, or at least no longer advertised that way.† But if you widen the definition just a bit, jingles have been with us for a very long time. When I started trying to do some research into the history of jingles, I hit a sort of roadblock. Every online source I found seemed to talk about the Wheaties jingle of 1926

★ In current parlance it's an audio logo. Jingles are now defined as having words.

† The fact that I can still remember all the cigarette ads from the 1960s indicates that the millions of dollars poured into their radio advertising budget was money well spent, at least from a brand recognition standpoint.

being the first jingle used in broadcast (radio, of course). All of them seemed to be either a paraphrase or in many cases a direct copy of the article on jingles from Wikipedia. The only other types of jingles mentioned seemed to be radio station ID jingles. All of them described jingles in exactly the same way, as a short tune used in advertising, a form of sonic branding. But really, a jingle is not just a way to sell you a thing—it's about linking an object with an emotion and transferring it to your long-term memory without requiring you to pay attention to it.

A successful jingle is an advertising tool based on the use of basic psychological and neurological principles. The thing is, advertising emerged long before either of these fields did.* Ancient wall paintings in Pompeii described things for sale; professionals in the Middle Ages put up large logo-like carvings indicating where you could get your horse shod, bled, or made into horseburgers; print advertising showed up in the first newspapers in the 1700s, using large letters in different fonts to grab your attention. Even in the twentieth century, people had the Oscar Mayer wiener song stuck in their heads long before the advertisers hired neural marketing researchers with access to fMRI machines. This is because over the eighty years or so that we've been able to broadcast sound into our environment, media people have learned the basic rules needed to get the job done.

A jingle requires five things to be successful: (1) it must be short enough to fit into your short-term memory, which means between four and seven notes or other elements; (2) it must be at a tempo that allows other sensory and motor systems to form correlations with the sound (i.e., you can hum it or tap your fingers to it); (3) it must be different enough from other things

---

* And not just in humans—remember that even frogs advertise.

you may hear so you can identify it; (4) it must be presented in the context of some *thing*, whether the thing is a specific object or a brand; and (5) it must use all these other aspects to create an emotional response. In other words, it requires frequency discrimination, temporal identification and integration, multisensory convergence, association, and connectivity with the emotional processing regions of the brain—all the elements that go into auditory processing and identification. But it has to go further than that. Hearing a jingle once does very little, so it must be repeated often enough to be transferred from short-term to long-term memory. If you hear it repeated a lot, a jingle will be logged by your brain as noise, which means both that you recognize it automatically and that you will start to find it profoundly annoying. At the same time, the presentation of the jingle must be strictly associated with a product so that you form a cognitive association with that product *and* provide an emotional context, so that when you recall the product, the emotion associated with the jingle comes along with it.

In short, the ideal jingle is an *earworm*, a piece of music that you can't get out of your head. For several years I've been working on an outside-the-lab bit of research I call the Earworm Project, in which I've been trying to figure out what makes an earworm and how to make the perfect one. There are numerous examples of good earworms or jingles—Walter Werzowa's Intel audio logo and Lance's T-Mobile ringtone are the two best-known examples. Lance described his process to me. The original T-Mobile logo (back when the company was known only as Deutsche Telekom) consisted of five boxes: three gray ones and then a pink one elevated out of the line, followed by another gray one. Five visual elements, with one different. Lance basically came up with a six-note logo formed of three notes the same, then one raised a third, and finally a return to the first tone, with the sixth just as a resonance fading away at the end. The simplified version is only five notes but they form an auditory match for the visual logo. Limiting it to five tones keeps it within the limitations of short-term memory, so it is easy to remember, and by using multisensory convergence (matching of sound and vision) you had two sensory systems coding the jingle or sound logo. One of the most powerful elements for integrating any sensory object is multisensory convergence— your brain codes objects that share characteristics, whether by color, nearness, or tempo, as a single object. (This is the base of auditory stream segregation; it works even better across modalities.) Multisensory convergence and coherence of stimuli seem to underlie what is called in neuroscience the *binding problem*— how the brain manages to take all the low-level sensory elements and form them into objects.

But it's not limited to just vision and sound. My wife, a sound and biomimetic artist, is also a former ballerina, and she pointed out that what always helped her remember music was the ability to move to it, adding tactile and proprioceptive (muscle position) feedback to the sound. Try this for yourself. Think of the first tune that you like or jingle that you remember, and I'll bet you that you can tap your toes or drum your fingers to it easily. So we expand the timing of a successful jingle to include not just the millisecond speeds of auditory recognition and the several-hundred-millisecond speeds of visual processing but also the slightly slower-than-vision rate of motor output.

One question is how quickly you can form an association between a sensory stimulus and an emotional state. It depends on a number of factors, including whether you are trying to form an instant association of emotion and sound or attempting to condition a listener to accept the sound as a substitute for what would normally trigger such an emotion. Certain sounds, such as a sudden roar—consisting of a loud fast-onset sound continued with lots of low frequencies with inharmonic, fingernail-on-the-blackboard components—are scary in and of themselves. While in most cases, someone trying to sell you something wouldn't want to scare potential customers off, I do remember a local haunted house radio ad that used a five-note minor-key jingle played on a detuned bass pipe organ that was pretty effective until they added the cartoonish scream at the end.

But since you can get a strong positive emotional response to sounds in less than a second, a brief series of sounds should be sufficient to give you a useful emotional association. This is one of the underlying bases of a type of simple learning and was first characterized by Pavlov in his early work in classical conditioning, which, as we've seen, involved conditioning dogs to associate the sound of a bell with food. The model most often used to describe the operation of classical conditioning is called the Rescorla–Wagner model, and while it was developed and tested on rats and pigeons, its basic precepts hold up on any organism complex enough to figure out how to chain more than two actions together to reach a goal (so, for example, this is beyond the capabilities of nematodes).

The problem is that advertisers rarely have the opportunity to shove a plate of their product right in front of you within half a second of hearing their jingle, as would be required with simple classical conditioning. So they rely on principles of learning to help you remember their product, form a positive association with it, and go and buy it later that day or week or year. This theory, called *operant conditioning*, is different from classical conditioning in that the learner has to modify his or her behavior to actually *do* something to earn its reward—such as spend money or go to that resort timeshare sales presentation.

And it's here that sound starts shifting away from something that enhances your suspension of disbelief, your enjoyment of a show, your emotional involvement with a created world. Here at the junction where people are beginning to use sound to deliberately shift how your brain and mind respond, with the goal of getting you to do something, we move from simple entertainment to the world of brain hacking.

Chapter 8

HACKING YOUR BRAIN
THROUGH YOUR EARS

A s you walk down the dark alleyway from the theater, the slightest noise makes you turn your head and startle. When the volume doubles as an ad comes on during a commercial break, you instinctively press the mute button. When the radio plays the song you heard when you had your first kiss, you stop what you're doing as brief memories well up without words.

Sound affects us in ways of which we are not aware. It changes our emotions. Changes our attention. Changes our memory, heart rate, desires, response to the opposite sex. Sounds like . . . is it . . . *mind control*? (Cue the threatening, *Manchurian Candidate*-esque soundtrack, which sort of proves the point). Well, yes, of course it is. The ability of the film score composer John Williams to manipulate the emotions of millions of people is why he's paid the big bucks. But there is really little to fear from this kind of mind control: music and sound work on preconscious levels, and a little understanding of the underlying mechanisms can explain how you (and others) can and do hack your mind through your ears—without the problems inherent in pushing the Q-tip in too far.

First of all, what is "brain hacking"? The first image most of us conjure comes from watching really bad movies and usually involves someone ending up with a light-bulb-studded colander on his head while someone else in a lab coat is screaming, "Fools! My creation shall destroy you all!" Then the switch gets thrown, and after the de rigueur exploding control panel and shower of sparks (because no one in science has ever heard of circuit breakers or fuses) the newly transformed creature develops mental powers that let him turn invisible or communicate with some hidden dimension where the electrical wiring is adequate. A more sedate version (and hence much less fun) can be found in consumer electronics catalogs where you can buy "mind/brain machines." These devices are supposed to induce altered states of consciousness ranging from pedestrian "meditative states" to my favorite, "synchronization with the electrical resonance with the earth itself," which sounds like what Nikola Tesla would buy if he had invented the Home Shopping Network before alternating current. You can even download sound files to "hack your brain with your iPod." It all sounds very exotic, and most of the vendors for these materials throw in all kinds of intricate-sounding "brainwave frequency ranges" with sexy Greek-letter nomenclature and more decimal points than you can shake your calculator at.

But brain hacking really just involves taking the research discussed in the previous chapters out of the lab and putting it to use in the real world. Brain hacking can be as simple as changing the virtual position of a sound from one ear to the other in time with basic rhythms of the brain to induce a change of mood. Or it can require complicated filtering and modulatory post-processing steps to try to get specific psychological or physiological effects. There are almost limitless possibilities for application of auditory neuroscience and psychology to the sounds we hear every day. These can include creating video games that make you feel more excited as you go up levels or actually feel ill as you lose points; composing film or video scores that use specific algorithms embedded in the soundtrack to manipulate viewers' emotions, without needing a full orchestral arrangement; using attention-grabbing sounds to shift a driver's attention to a blinking dashboard light or to people about to be run over; and what you would think would be a real money-maker, making advertisements "stickier"—creating an audio logo that's impossible to get out of your head. The question is, can these techniques work consistently enough to form the basis for modern marketing?

To see what sonic brain hacks can and can't do, we have to examine some specific applications. Brain hacking, aka ways of altering states of consciousness, can be broken down into two major types: (1) ones that induce global changes and generally change your brain's overall state by increasing your arousal, and (2) ones that modify specific elements of your mental state without inducing overall cognitive changes. The two seem very different on the surface, but either can be brought about by using a few simple rules for controlling sensory input. And if you are the hackee, understanding these rules can either enhance your experience or help you fend off the effects when you find out someone is using such manipulations on you.

Let's take a look at the simplest approaches first. What's interesting here is that two of the most effective techniques seem diametrically opposite in their nature: either limiting the sounds your listener hears or increasing them to the point of being overwhelming.

Brain hacking using limited sounds is probably the simplest:

just cut out the noise. Take noise-cancelling headphones, for instance; they may not seem like a brain hack, but they assuredly are. They shut out your normal auditory environment so you can focus on the internal one, usually your music or your audiobook. Think of it this way: if you're walking down the street or sitting on a plane wearing noise-cancelling headphones, you are cutting out all the environmental signals that your subconscious normally processes to help you orient yourself without thinking about it. This is a boon in a plane, as it's not likely that hearing a change in the sound of the engines is a good reason to leap up and storm the cockpit to save the passengers on your flight. But wearing such a headset while walking or running on the street is more likely to make you miss something important, like that SUV making a right-hand turn from the left-hand lane. Your exercise in auditory damping thus yields results closer to brain squishing. You would have done less damage with the Q-tip.

But if you're still not convinced that damping the noise is a serious form of brain hacking, call around to a local university and find someone who does auditory research, preferably someone who works with bats. Bat scientists are a bit weird. They will be happy to talk to you, as they don't get out much in the daylight and a lot of their social interaction involves math or discussions of things such as how a pallid bat can hear a scorpion fart from five meters away. While you may not think of these as the best pickup lines, these scientists often have cool toys and specialized gear for playing with animals whose hearing is several orders of magnitude better than yours, as we discussed earlier. More to the point, they probably have an anechoic room for the bats to play in. And an anechoic room will show you how warped true silence can be for normally chatty humans.

You walk in and see that the walls, the ceiling, and occasion-ally the floor are lined with egg-crate-shaped foam to absorb sound or reflect it to another absorbing surface. Close the door and it gets quiet. Real quiet. After about two minutes, you can't help but say, "Damn, it's quiet in here," just to hear something. But the lack of echoes and reverberation muffles and damps your own voice. Make a sound and it gets swallowed up instantly. And your brain tells you that is just *wrong*. Some people begin having anxiety issues within a minute or two, but most people can hold out for another minute or two before they start hearing a faint hiss. A really good anechoic room is so quiet that your ears are suddenly able to show off their stuff and you start hear-ing the air molecules banging around. *Hissssssssssss*. And then you start getting the "I'm stuck in a Vincent Price horror film" feeling, as above the hiss comes a gentle and quiet *lub-dub, lub-dub, lub-dub*. It's your own heartbeat. And that's all that you can hear. That's when most people leave, or start singing show tunes. So there is the simplest of auditory brain hacks—just remove all the extraneous noise of echoes, reverberation, and the susurra-tion of background voices and noises and you are definitely in a heightened state of awareness, trying to figure out what's wrong, what's missing. It's one of the best demonstrations of how ex-pectant, if not dependent, our brain is on normal, unattended background stuff. Take it away, pare your auditory world down to the bare bones of basic signals, and your whole mind shortly starts twitching—and waiting for the giant bat to come and de-vour the farting scorpion before your very eyes.

If being driven insane by the sound of your own heartbeat isn't your thing, perhaps you'd prefer to hear a *lot* of sound. Un-der some circumstances we all like loud input—turning up your portable music player, blaring your stereo or TV, or going to a movie where the explosions seem to constitute 50 percent of the soundtrack. But what underlies this drive for loudness? And how can it be used or misused? The answer is pretty simple and clearly defines high volume as a brain hack: loud noises activate your sympathetic nervous system. Your sympathetic nervous system is the control system for the three f's that drive a lot of vertebrates' lifestyles: fight, flight and fffff-igure it out.

Very loud sounds, especially ones with sudden onsets, are treated differently. First of all, loud noises activate your whole inner ear, not just the cochlea, the part that normally detects and codes sound. If a sound is loud enough, it will trigger non-cochlear parts of your inner ear such as the sacculus, an organ normally devoted to balance that picks up low-frequency vibrations and lets us know which way is up. If there is a sudden loud bang, even the most acoustically OCD person is not really concerned with its frequency content, its possible linguistic meaning, or whether it is consonant or dissonant. He is mostly relieved by the fact that whatever made the loud sound didn't land on him. A loud, fast sound makes you respond with a highly stereotyped and very fast behavior (usually only involving three neurons): the sound activates the sacculus (normally a gravity sensor), which triggers high-speed motor pathways involved in posture control, making you hunch your head between your shoulders and jump a little. In short, you startle. But beyond the quick motor response, the loud noise also causes both auditory and vestibular signals to pass from your brain stem in the other direction, along a slightly slower route to regions that increase arousal and alertness, just in case the anvil that just landed next to you was merely the first of many.

Is a sudden startling sound a brain hack? Sure, if you use it right. Take a movie with not much happening. All is quiet. The spaceship is cruising with all systems green or the family is happily having dinner. And then—*boom*. A well-done, well-placed sudden sound and you startle right out of your seat as the Thing pops out of someone's chest onto the spaceship lunch table or the jet engine comes through the dining room ceiling, still attached to much of the jet. The audience's heart is racing, seats are damp. I think that qualifies.

But you can argue that though a loud sound may startle you, it wears off easily. Startles are triggered only once or twice and then the person habituates to the signal. Yell "boo" at a baby a couple of times, and by the third time, the kid just looks at you like you're crazy or cries or both and the diaper is full and the baby-scaring fun is gone. Or a scene in a movie where the monster keeps on roaring as it pops through everyone's chest becomes a joke, not a terror-inducing moment. A noise-induced startle is a good means of showing how signals from your ears can trigger a short-term arousal response, but what about loud continuous or repetitive sounds?

At a physiological level, as mentioned earlier, loud is bad. Your ears and your brain respond to excessive volume, whether startling or chronic, by trying to get control over the signal or, failing that, trying to get control over your own behavior. If you aren't making the loud noise, you try to figure out where it's coming from and move away, plug your ears with your fingers, or just bang on the ceiling with your broom to get your neighbor to shut the hell up. So why do we sometimes immerse ourselves in loudness? For the same reason that some people BASE jump or ski or drive too fast—because activating the fight-or-flight system gives you a rush, a temporary surge of excitatory neurotransmitters such as epinephrine and dopamine which light up the arousal parts of your brain like a neon-bedecked Christmas tree at a rave.

If the loudness is not under your control, you get a different set of behavioral effects as you try to bring your sympathetic nervous system under control (presuming, of course, you can't just leave or turn down the volume). Your brain has a specific mechanism for dealing with sensations that are under your control—it's called an efference copy. It's the reason you can't tickle yourself and why you usually won't get carsick if you're the one driving. It's an automatic program that links your brain's executive decision-making centers with your perceptual centers via a mechanism called *motor-induced suppression*. One of the more familiar times it gets called up is during speech production, because if you're yelling to get a friend's attention across a room, it's going to be *really* loud inside your head.

To prevent deafening yourself, this reflexive system drops the gain (or relative loudness) and decreases your auditory sensitivity to your own sounds, but it can get activated even when you engage in something non-speech-based, such as reaching for the volume knob. Let's say you're listening to your favorite song and you decide it's a good enough track that it really deserves to make the walls pulse and your chest resonate. Your brain makes a brief plan, not only of the motor functions it will have to carry out to reach over and turn the volume up but also of what it expects will happen (things will get louder). This feedforward command sends information to your auditory system that things are about to get loud and actually reduces the sensitivity of your brain to incoming sound. You can tell it's louder, but it will damp out the what-the-hell-was-that aspect of someone else making an unexpectedly loud noise in your area.

But what happens if you're not the one in control of the

volume? How many times have you been watching a TV program or a video on the Internet and suddenly the ad comes on
twice as loud as what you were watching? This is deliberate—
it's a way of getting your attention and increasing your arousal
by activating your sympathetic nervous system. But while it
may have been a clever trick long ago, when radio and broadcast TV still ruled the Earth, its overuse has created its own
downfall: it spurred the rise of TiVo, which allows viewers to
skip ads entirely; the recent signing into law of the Commercial
Advertisement Loudness Mitigation (CALM) Act, which requires broadcasters make sure that ads are not louder than the
programs in which they are presented; and the adoption of a
mute button on every remote control device out there, returning
control over your auditory world to you. This is, unfortunately,
characteristic of the way media often try to use psychological
"tricks" to increase consumer name recognition and attention.
While the technique uses one basic perceptual principle—
increasing attention and arousal through loudness—it ignores
an even more powerful one, habituation. So instead of increasing your memory of a brand and forming associations between
it and the desired sympathetic f's, instead the noise ends up
making you annoyed with the ad and hence with the product,
and using the flight f option to avoid it.

But what about when the environment you choose to be in
is inescapably loud? You've no doubt run into this scenario. You
walk into a bar where the din is so loud you can't even hold a
conversation without yelling into someone's ear. You go into a
store in a mall, particularly those that cater to a younger crowd,
and suddenly you're bombarded with 90 dB music levels that
would make an OSHA inspector reach for her earplugs even
before writing up a citation. Or you move through the quiet

entrance lobby of a casino into the gaming room, where the noise from hundreds of machines, bells, alarms, and other come-on sounds combines with the blinking lights to create an environment where you can't hear yourself think. And that's the point. Thinking can lead to rational, measured choices,* and frankly, whether you're in a bar, a pricey store, or a gambling casino, rational, thoughtful choices are not going to help such an establishment's bottom line. The sound levels in these places are not the result of poor design or acoustic accident. The point of such overwhelming sound is to increase your arousal, to activate your sympathetic nervous system. But since you have elected to go into this environment and have no way to lower the sound level, you will do what almost any animal will do when faced with an inescapable stressor: find a way to control it. In one study, rats were exposed to the choice between an unpredictable electric shock and pressing a switch that triggers a predictable one; they learned to choose the latter even though it meant pressing a switch that hurt them. And multiple studies have demonstrated that buying things or gambling—"acquiring controllable resources," if you prefer the literature's terminology—is a common strategy for exerting control (real or imagined) over stressful situations.

But it's not an automatic "buy stuff" button. If I, at the tender age of fifty-one, am in a mall and pass an Abercrombie and Fitch, where the corporate-set level for music (according to one news report) is at 90 dB, I'm going to walk away from the entrance, not even exposing my eardrums to construction-site-level come-on tunes. But younger people are accustomed to and often seek out louder noise and music levels (which is why

---

* If you are a very good thinker or very lucky.

younger drivers eagerly buy car mods that reduce muffler effectiveness to make cars louder). So this strategy is targeted to a specific demographic, the younger, more acoustically arousable purchaser. Similarly, in a casino, studies examining the effects of noise on gambling showed an interesting demographic effect. Casual gamblers showed an increased tendency to gamble more when exposed to loud noises, whereas those who were identified as having a serious gambling problem tended to bet less. One hypothesis to explain the difference is that while both groups are being subjected to increased arousal from the loud noise, the casual gamblers associate arousal with winning and bet more, while the problem gamblers associate it with losing and bet less. These two examples point out a very important point about sound-induced arousal: it varies based on who is listening, and thus affects the likely outcome of acting on the arousal. The targeted use of loudness, from louder ads to deafening store music, is an important but often misused marketing and sales tool.

Turning the volume way down or way up is the simplest form of brain hack, but how useful is it? You're not likely to have an anechoic room handy at all hours and even the best noise-cancelling headphones drop ambient sound by only 45 dB or so (earplugs are even worse, at about 25 dB). And while too many of us, especially kids, turn up the volume too high or get stuck in noisy bars or casinos, we can and do ultimately escape—if not by leaving, then by the inevitable loss of hearing a few years down the line.

So how can you induce a global brain hack without access to a specialized facility or risking deafness? By realizing that volume is only a small part of the auditory world.

Another critical part is time. By manipulating the timing of sounds, you can force large chunks of your brain that normally do their own thing into artificial synchrony.

The brain is used to dealing with a large number of simultaneous and asynchronous inputs—it's a noisy world out there, even though it's the one you evolved and developed to deal with. If you can eliminate the noise and overload your brain with one type of input, this sensory focus will do some very different things to your mind. This is the basis for a lot of mental states that go by a variety of names—alpha state, meditation, trance induction—but all it boils down to Global Brain Hacking Strategy Number 3: limit the distractions to intensify your focus. This is probably the type of brain hacking that you've probably heard the most about, because it is the basis of hypnosis and meditation.

Hypnosis and meditation seem like they spring from opposite types of input: hypnosis is usually brought about by surrendering your executive functions, your decision-making ability and attention, to another, while meditation is usually self-induced, by blocking environmental distracters from your attention and using some unifying stimulus from inside or out to help you achieve a more narrow state of focus than you usually aspire to before coffee. But despite the divergent approaches, both change your attentional state by limiting the amount of your brain you devote to processing the sounds of someone else's cell phone conversation or that dripping faucet and free up those areas for other things, such as shifting or increasing attention, something very important for brain hacks.

What kind of stimuli can induce a person to stop paying attention to anything but a single focused input? Words spoken in a calm, rhythmic voice can act as an important focusing element, but often hypnotists will use a regularly flashing light or the rhythmic ticking sound of a metronome to help the induction of a trance state. Why? Because attention is about shifting neural resources onto an object of interest, and the more attention you pay to an object, the more you pull neural resources from distracters. By choosing to pay attention to a specific sensory input provided at the correct rate, you start swamping local variations in parts of your brain that would normally slide off to wonder if you turned off the lights before leaving or what you're going to have for dinner or even why the guy speaking to you so calmly would want you to squawk like a chicken. You are narrowing your sensory world to focus on overwhelming and synchronized sensory input. And as your normal waking brain is strongly dependent on differences in both the quality and timing of input, overwhelming it with a narrowly focused signal, ticking away while only a single voice speaks quietly and soothingly at the right rhythms in your ear to maintain this odd mental state, is why you see nothing wrong with squawking like a chicken after all.

But what are the "right rhythms"? Your brain has intrinsic rhythms, ranging from neurohormonal changes over periods of months to a single neuron changing its activity state in milliseconds or less. The most commonly referred to rhythms are the ones that involve large swaths of your brain hemispheres and are the ones usually reported by electroencephalography. While electrophysiological recordings of living brains actually go back as far as the 1840s, it was in 1920 that Hans Berger developed the EEG recording device, which was capable of noninvasively recording electrical signals from living brains. Modern EEGs are quite spatially selective, able to record from dozens or even hundreds of individual electrodes placed on the head, and are capable of extracting information about the timing of neural signals down to the millisecond level. However, they all suffer from the same limitation—they are recording very weak signals (thousandths of a volt or less) through a very good insulator (the brain's protective membranes and your skull, scalp, and hair). This means that the only signal that makes it through is a really a summation of thousands to millions of individual neuronal signals. You can't analyze the response of a single neuron or even a local circuit, but with proper electrode placement, you can measure the responses of populations of cells that act in concert, and hence get a global view of how big volumes of the brain (on the order of millimeters or centimeters) respond to specific stimuli such as flashing lights or sounds, known as "evoked potentials."

But even when you are not collecting evoked potentials, the free-running brain is never silent (unless it's dead), nor is it chaotic (unless the subject is having a major seizure). And while an untrained person looking at a raw EEG tracing sees what looks like almost random rises and falls, buried within the mass electrical response of the brain there are five major rhythms that underlie global functions of the human cortex, all combined, but each changing under different physiological or cognitive conditions. The theta rhythm is the slowest at 4–8 Hz and seems to arise, at least in part, from the hippocampus during memory processing. The alpha rhythm, cycling 6 to 12 times per second (6–12 Hz), is generated by connections among different parts of the cortex and between the cortex and thalamus, the brain's relay center. It is often subdivided into the lower-1 alpha band (6–8 Hz), the presence of which indicates alertness; the lower-2 or posterior alpha band (8–10 Hz), seen during changes in attentional states; and the upper alpha band (10–12 Hz) which is often triggered by external events and language-based memory tasks. The beta rhythm (20 Hz) is generated in the motor cortex, which controls voluntary movement, and is usually only seen shortly after the subject stops moving (sort of an "off" or "end program" signal). The gamma rhythm is the fastest at 40 Hz and is one of the more interesting and controversial of the brain's major rhythms.

Several studies have described the presence of a gamma "wave" traveling from the front to the back of the brain, sweeping across much of the cortex. This has led to the hypothesis (still unproven) that the gamma band may be involved in binding together all of the individual sensory inputs and feedback loops that let you perceive the world as a consistent place that doesn't wobble about between blue, high-pitched, nasty-smelling, and too hot.

These rhythms are part of the basic infrastructure of the working brain. The presence of these rhythms in EEG traces are evidence of the coordinated firing of millions of interconnected neurons, underlying functions critical enough for the brain to devote large proportions of its processing power to them. Access to these functions is an opportunity for brain hacking—and for marketing of devices that use such access with highly varying degrees of effectiveness.

Trance bells, mind-brain machines, neural feedback devices, iPod brain hacks—even the most basic online search will lead to dozens of hardware or software products claiming to unlock the power of your mind using sounds beeping and lights flashing at one or more of the rhythms mentioned above. While most of them are based on less real science than a 1950s Godzilla movie, the funny thing is that a lot of them will work anyway because of user expectation. If you are out shopping for a brain-mind machine to let you supercharge your executive powers, well, you're already halfway convinced, and you've probably already spent money on sillier things, such as sleep-learning tapes to improve your dog's self-esteem. (No, I'm not making that up. After all, you do have to worry about the self-esteem of a dog that lets its master put headphones on it. Perhaps after a while it will dream of chasing really big rabbits.) In fact, it actually is possible to induce changes in mental states by correct manipulation of sensory information at rates that use these basic rhythms.

At first it might seem easiest to just play a tone or a noise at a specific rate, say the 8–10 Hz posterior alpha rhythm, through a pair of headphones and wait for a soothing sensation to take over. Unfortunately, it's more likely to become rapidly annoying or boring enough to drive you to think. This is because of a number of factors that we've mentioned before, including adaptation, habituation, and the fact that playing a single tone to both ears simultaneously will use only a very small percentage of your auditory processing power. Whether the sound uses the posterior alpha rhythm or not, it's just going to be a rapid pinging in your ear, like a microwave that really wants your attention. This, it seems almost needless to say, is super-annoying and probably will not hypnotize you.

To get massive amounts of your brain entrained to a single rhythm, you need complicated input from a number of sources all acting in concert. One method is to use binaural beating. It's a remarkably simple way to drag more of the brain into processing sound at desired rates. If you play, say, 440 Hz in your left ear but 444 Hz into your right ear, you will hear not two separate tones irritatingly close in pitch but rather a single tone that seems to be modulating four times per second. This is due to the action of the superior olive, an auditory nucleus in your brain stem that lets you figure out the position of sound in space based on relative amplitude and/or on time differences between your two ears.

The superior olive is the first place in your brain that receives input from both of your ears, and it carries out mathematical analyses to determine where the sound is coming from in space. If both ears are presented with the same tone, the superior olive passes the information to the rest of your brain and you perceive a single sound in a fixed position, such as in the middle of the speakers in an open room, or in the middle of your head if you're wearing headphones. However, if there is a slight difference in frequency between the two ears, on the order of a few Hz, you get the sensation of the sound moving back and forth between your ears at the rate of the difference between the two tones. If the frequency difference is a bit larger, around 4–12 Hz, what you hear is a single tone that seems to change in amplitude at the modulation rate. With an even greater difference (typically on the order of 20-40 Hz), you will in fact hear two separate tones and be convinced all is well; your head will be mercifully unthrobbing. So binaural beating can be a simple but effective way to entrain your brain from your brain stem up to your cortex as long as you are using simple tones and relatively low modulation rates, such as those in the alpha, theta, and delta rhythms.

If you listen to binaural beating tones, you can get some meditative or attentional effects, depending on your level of expectation and preparation for these states. But listening to single tones warbling over and over for twenty minutes can test the patience of even the most dogged pursuers of altered mind states. Most people who have tried it agree that it has an effect but they never want to do it again, preferring to have three or four mojitos to achieve an equally altered state with more potential for dancing. So some of the more effective brain hacks use complex sounds composed of multiple frequencies (music, speech, car crashes), which in and of itself devotes more brain processing to the task, and turn up the power more by modulating not just the amplitude but the relative position as well—what sound engineers call "panning." You've heard numerous examples of this technique, particularly in 1960s and 1970s classic rock, such as the opening to Pink Floyd's "Welcome to the Machine," where the engine-like sounds appear to move relatively slowly from one speaker (or ear) to the other by smoothly shifting the relative volume of the sounds from left to right.

But the superior olive uses amplitude differences only for relatively higher-frequency sounds (above about 1,500 Hz for humans—think of the shriek made by a small child who has just found a bug in her milk). For sounds made up of frequencies below this, the superior olive relies on differences in fine timing or phase differences between the two sounds. So if you use a complex musical piece and take care to synchronize the phase differences for low frequencies and amplitude differences for high frequencies, you can get a frighteningly realistic sense of apparent motion from the sound even just using stereo speakers or headphones. And it's also how you can remix almost any piece of music (aside from drum solos) to evoke mind-altering states. Here's where it starts to get fun.

What if you *don't* synchronize the modulation rates? What if you make the high-frequency sounds go at one rate and the low-frequency sounds go at another rate and constantly change the rates in between? Then you potentially have a very serious, effective, and fun (for some people) or awful (for other people) brain hack.

Here's an example. In my misspent youth, I got a phone call

from my old friend, Lance Massey, who asked, "What is psychophysics?" I gave him a very long lecture about sensation and perception. Then he asked the question that still is causing us trouble: "Does this mean we can do mind control with music?" My answer, based on years of study and experimentation, was, "I dunno. Let's try it." And so we became the frontmen for a very unusual band.

Lance composed a series of ambient musical pieces to which we applied embedded modulation rates based on best stimulation frequencies for different parts of the brain, to try to create specific psychological effects in a listener. The idea we had was to use music to elicit neuronal responses that targeted specific areas in the brain. It's similar to how modulations of a carrier wave transmit information via radio. You're not that interested in the carrier itself—you are just using it to carry the modulations to a receiver where it gets converted into a useful signal for the listener. This kind of sonic algorithm is several steps more complicated than simple binaural beating. The idea is to modulate amplitude, frequency, and phase characteristics of almost any musical or other complex acoustic signal at frequencies that best stimulate the parts of the brain that do certain things, such as induce emotional responses, change heart rate or blood pressure, make the listener feel as if he or she is moving, or just change the listener's attentional states.

One of our first efforts was to use phase and amplitude changes that would make the music seem to orbit a listener's head. At our first public concert, in a little hole-in-the-wall club in lower Manhattan, we noticed people slowly wobbling around in their seats when the piece came on—and when one guy fell out of his seat as a big virtual motion shift kicked in at one point in the music, we knew we were on to something. Just by using this orbital position sonic algorithm, a simple ambient music piece began to do something weird: control how people thought they were moving in space. So we decided to ramp up the evil meter to 11 by slowly changing the modulation rates for different frequency bands. We thought a piece that used this algorithm would make people feel like they were moving but confuse them when some of the sound elements seemed to move one way and some the other. We lovingly titled this "The Vertigo Tour" and unleashed it as a track on our CD.

Despite the fact that I knew I was torturing my listeners, I begged and pleaded for feedback from anyone who got a copy. It broke down in an interesting fashion. After a few minutes of listening, about one-third of the listeners felt like they were moving when the amplitude and phase were synchronized, another third thought the music was moving through the sound field, and the final third got violently ill. Yay science. We had figured out how to induce auditory motion sickness. Unfortunately, the chair of my department at the time, a dedicated audiophile whose sound system was worth more than my salary, fell into the third category. The day after I handed him the pre-release CD, he emerged from the elevator while I was trying to get caffeinated, grabbed my arm, and said, "Your music made me sick!" "Then I guess we win!" I said, not too brightly. After all, just using relatively simple modulation rates, carefully applied in a complex musical piece, I managed to hack my department chair's brain and make him almost throw up all over his expensive stereo system.

Luckily, I found another position relatively quickly, as even I realized that the financial possibilities of an acoustic vomit stimulator were rather limited. But it proved to me that acoustic brain hacking could have profound effects on the listener, even at a physiological level. And while inducing vomiting is not really among the top ten effects you'd like to get from your listening audience, it demonstrated that it might be possible to start getting targeted, specific responses to manipulated sounds. So what about using sound to get effects that somebody might actually want to induce in a listener?

Music and sound have always been used as a means for inducing emotional and other responses in audiences, and they do it under your cognitive radar. Movies don't put up big signs saying "shriek in fear" right before the monster taps the hero on the shoulder, and asks him to get off its tail, at least not in the same way as prompt signs tell a TV studio audience to applaud. Instead, the slithering sound, the sudden silence, the overwhelming roar go from the ears right to the parts of the brain that modulate emotional response via the sympathetic nervous system, and you find yourself scattering your popcorn. So when word first started getting out about our concert/experiments, we were approached by a British metal band called Thrush who told us they were writing a song about brutality in the legal system and wanted us to modify the song so that anyone who heard it would shriek with terror. This is not the kind of thing you normally think of as a desired outcome for listening to music, but that's metal bands for you, and we accepted the challenge. And once again, acoustic science gave us the answer by applying sound to the emotional connections in the brain.

Working feverishly in the dark, secluded Bavarian castle of my Long Island apartment, I made a filter that would take any sound and throw our old friend the pseudo-random fingernails-on-a-blackboard envelope over it, then applied it to the metal band's song. When we played it back for final re-recording, the studio engineer ran out of the room screaming and refused to work with us ever again. Thus is success measured in the world of neurosensory algorithmic brain hacking. At least this time no one had to mop up.

So by reverse-engineering a sound with a known emotional and physiological effect using relatively simple algorithms, it is possible to capitalize on intrinsic brain properties to get targeted perceptual effects just using music as a carrier wave. But what about doing something both less disturbing and more specific? Something with a very specific outcome that didn't involve vomiting or running away screaming?

A friend of Lance's was the happy father of two rather hyperactive children, and as most parents do, he was complaining about the impossibility of getting the kids to sleep without recourse to medication or a baseball bat lovingly wielded. He had heard about some of our work and asked us very straightforwardly whether we could come up with some way to make his kids calm down enough in the evening to go to sleep.

Now, sleep is a really complicated phenomenon. We don't really know what it's for, how it works, or why it's so prevalent in any organism with more than two ganglia to rub together, but having worked in a chronobiology lab for several years, I knew about one of the biggest problems involved in all long-distance trucking—that of falling asleep at the wheel. Driving anything from a train to a car should be able to give you enough motivation to stay alert so that you are not awakened by hideous crashing sounds. But falling asleep on long hauls is very common—so common that a lot of grant money is spent on industrial research into monitoring driver alertness. You can even find little electronic head-mounted devices that shriek unpleasantly at you if your head drops below a certain angle, as happens with most people who nod off.

The underlying principle behind falling asleep at the wheel is rather surprising. It's not exhaustion or attentional issues, although these can contribute. It's actually due to the fact that your vestibular system, the balance part of your inner ear, is extensively connected to your arousal centers as well as centers that affect non-voluntary things such as salivation and gastric control. You've probably had some experience with this if you've ever gotten on a boat in high seas or on a really high-speed roller coaster—your vision can't keep up with your inner ear's signals and you get sensory dissonance, which leads to motion sickness. But nauseogenic motion sickness (the type that makes you want to redeposit your lunch) is only one form. Low-amplitude, low-frequency pseudo-random vibrations, the kind experienced while driving even on a relatively smooth road, yields another form of motion sickness called Sopite syndrome, which leads to extreme tiredness and sleepiness, no matter how important it is that you stay awake. Few people have heard of this, although the effect itself is relatively common knowledge among parents of infants—it's the reason you can rock your child to sleep. The low-amplitude, low-frequency oscillations lull the child and she drifts off. My parents used it by depositing me into the backseat of my father's old suspension-challenged VW bug and driving me around a local dirt road for half an hour, which worked even better.

An interesting feature of Sopite syndrome is that it can affect passengers much more than drivers, and it is possible that our old friend the efference copy is responsible—if you are driving, you have at least some control over the car's motion, and the rocking and shaking have to go on for much longer to get to you. (The clearest example I ever had of this was on the second date with my fiancée-to-be, when after about an hour's drive she nodded off. I felt touched that she was so trusting of my driving ability, when in fact years of married travel experience later showed me that she was just really sensitive to Sopite-type stimulation. Romance and science are often at odds.)

So, weighted down with papers based on years of research at NASA, NIH and military organizations we set out to figure out how to make hyperactive kids sleep. But because we were doing this as a personal experiment (i.e., weren't being paid for it) and were pressed for time, rather than compose original music we decided to just mix our modulations into some old recordings of classical music tracks. This was an interesting challenge, as we needed to make subtle changes in amplitude at the right rates but not change the overall sound of the piece. After all, Dad was not telling the kids that he was selling their souls to science; he just was going to put on a CD of "boring" classical music in the hopes that they might calm down. This had been tried in the past with about as much success as you might think. The problem was that when we tested it on my very Sopite-syndrome-sensitive wife, there was no effect. So we stepped out of the realm of algorithm and into the real world of recorded sound. We placed a geophone, a very low-frequency microphone used to record earthquakes, in the back of my car and drove around for twenty minutes, then extracted the envelope from that sound, convolved it into three different classical pieces, and handed it off to the weary father with our fingers crossed.

The next day he told us that he'd tried the first track and it hadn't worked. Feh.

But the day after that he called us up and announced we were geniuses because both the kids had slid off to sleep within minutes after he put on the second track. We were pleased to have our genius recognized, but it was a bit confusing. The songs were about the same acoustic density and had similar keys, instruments, lengths, and volumes, and we had applied the same algorithms. Why did one track have no effect and the other two tracks worked so well that the disk ended up being called "the going-to-sleep CD" by one of the kids? For a reason that highlights one of the big problems of brain hacking: that all brains and all individual brains' experiences are different. It turns out that these particular hyperactive kids were enthralled with a thrash metal version of the first track that they had heard on the Internet, and it was in fact one of the things that made them bounce around like Tasmanian devils on speed. So while the algorithms were there and were clearly effective as sleep inducers in the other two tracks, the kids' previous experience with the basic musical structure of the first piece in a high-speed and exhilarating version had overridden the relatively subtle signals that the modulations provided. And as I've stated before, that's one of the problems (and, on the consumer side, comforting factors) with brain hacking of any sort: if it is based on statistical models of how the brain works (which is what neuroscience is largely based on), while it may work on most people, it isn't going to work on everyone.

So how can we use brain hacks in the many facets of our lives? I started playing with targeted auditory brain hacking in 2000, trying to figure out how it could be used legally. At the time we were the only ones really doing it. And our first idea was advertising. After all, advertising has been around since the first multicellular organism grew something showy to impress its swampmates about its fitness or its ability to defend its territory. It's been a principal part of human communication since the first exaggerated bone sculpture that showed a hyper-fertile female form ("Come to Ogg's cave for a good time. Mammoth parking out back"). And even in its more contemporary sense, advertising is about getting an emotional hook into potential consumers to convince them that they desperately need something that they never even knew existed until they saw or heard the ad, lest they be unattractive, unhireable, or just out of the loop. Without getting into a long discussion of the idea of consumer psychology and the new field of neuroeconomics, advertising, especially successful advertising, is based on fairly solid psychological ground. It's about making you buy stuff. And one of the key points about any purchase is that almost all purchases are emotional. When we decide to go out and buy a new music or media player, few of us really spend hours researching the latest electronics, checking specifications, comparing how flat the frequency spectrum of the headphones is, or whether its maximum volume violates OSHA or health guidelines. We buy it because it looks cool or our friends have it. Despite all the outrage over the recent introduction of a music player targeted to women (the difference was not that it preferentially lateralized speech to the right ear or had a slightly different frequency response, but rather it that only came in pink), that marketing decision had a reason behind it—it sold to the target demographic.

Advertising is based largely on coming up with catchy phrases and jingles to push a consumer toward an emotional need for an item that she could probably live without. The idea is to convince you that you really need the Omniglot 4000 Home Pork Puller with salad shooter because without it you will be unattractive, have questionable hygiene, and will probably die of restless eyebrow syndrome. But ads are supposed not only to convince you that you need something but to use psychological tools to help you remember the item's name and be hooked into the idea that only *that* particular brand will give you what you need. In short, the aim is to manipulate a consumer's emotions, memory and attention by getting the ads to stick with you, using basic psychological principles (repetition is the heart of memorization, sex sells, etc.). And if you doubt for one second that these basic principles work, well, how come you remember exactly what comes after "My bologna has a first name"?

But no one in the 1960s was using advanced neural algorithms to plot control of the consumer mind. They were just using what had worked in the past, and sometimes they got it just right. They set out to create catchy jingles, but by an accidental confluence of factors ranging from familiarity to musical key to rhythms, they wound up making earworms (or "auditory perseverates" in neurospeak)—musically linked phrases that are almost impossible to get out of your head. Earworms can be as simple as a nursery rhyme phrase stuck in your head for a few minutes or as complex as the opening four measures of Bruce Springsteen's "Thunder Road" cycling through your brain for five solid days until you are ready to nuke New Jersey to make sure that nothing like that ever again emerges from the Garden State.

There are people studying the bases for this phenomenon with an eye toward application in what is now being termed "neural advertising." So far the data show that it's not just rhythm or tonality or content. It appears that earworms may be based on a number of rhythmic activities in the brain that when synchronized make it very hard for the sufferer to break the loop. And the day that some advertising house finds the key to making the ultimate earworm that will never let you escape its brand-name looping goodness will be the day that there will probably be mass lynching of ad executives and neural advertisers all around the world. ("There are some things man was not meant to meddle in, my dear Dr. Frankensound.")

But how much of this talk of "neural advertising" is real? Is there really an industry out there that uses neuroscience and psychology data? And the bigger question is, does it work or is it the buzzword of the week? Yes, it is real, and it can work, but it usually doesn't, for some pretty basic reasons.

Let's look at a specific example of what you might try to do with auditory brain hacking and examine how well it might work. Imagine you're an advertising creative type and you have to sell a romantic product targeted for women. Men's and women's brains are different, and they hear differently as well, although spending a lot of time calculating the gender gap in loudness and frequency thresholds probably wouldn't show enough of a difference to be worth it at a practical level. But if you remember back to an earlier chapter, an awful lot of the auditory system is dedicated to detecting and differentiating between desirable and undesirable mates. Sound-based mate selection by females is not limited to frogs—just look at the demographics for Barry White fans and you will get my drift. A deep, low-pitched voice implies a big source, and in sound as in life, size does matter, especially to the subconscious bits of your mind.

But let's take it a step further. Size is highly correlated with pitch due to volume and mass of the vibrating surfaces that both make and detect sounds, and for us bipedal types, height is a big consideration in mate selection processes. So the sound designer can use a low-pitched sound or voice and also fool the ears into thinking that the sound is coming from a point higher than the listener.

This is tricky. While the superior olive can localize sound left to right (and front to back to a lesser degree), vertical sound cues are strongly dependent on the shape of your outer ears, with their little bumps and valleys. The trouble is, everyone's outer ears are slightly different. (There have even been a few forensic cases in which an ear print was used for identification purposes.) Your brain, having lived with your ears for quite a while, is able to interpret whether things come from above or below based on notches (frequency-limited reductions) and peaks (frequency-limited amplifications) in the sound. Luckily, sound localization, even in the vertical plane, is a well-studied area of auditory psychophysics. There have been a few studies demonstrating that if you create a filter that changes the position of a spectral notch from 6 to 8 kHz in wide-band noise, it will appear as if the sound is moving upward. So our sound designer, being a genius and having way too much software and time on his hands, takes a vocal recording saying the name of this romantic product, pitch-shifts it down to the Barry White zone, and embeds it in quiet sound that has an upward-sweeping high-frequency notch, and suddenly the target audience is hearing the name of the product in a very sexy voice coming from a very tall potential mate. Fire off the signal to a wide selection of sites in the brain, ranging from the auditory cortex to attention areas in the prefrontal cortex and to hypothalamic and preoptic regions controlling things ranging from pupil size to arousal, and voilà—you *may* have attracted the listener's attention and created a very positive, very sexy deep-seated response to something that was just a name.

If only it were that easy.

Your listener could be a man. Men often react to this kind of manipulated sound with irritation or annoyance. So they aren't going to buy this product as a gift. A woman may not be interested, even at a deeply wired level, in men. Or perhaps the listener's ears are such that she gets her vertical position information in the 8–10 kHz range. Or perhaps a case of Barry White albums fell out a window when she was a child and killed her kitten. Brain hacking will work on a statistically significant portion of the population, but people's brains are even more individualized than their fingerprints. And statistics are only valid for populations—you can't predict the success or failure of any given stimulus on any given individual.

But a "statistically significant portion of a population" represents a healthy demographic for marketers. This is probably why news and media outlets are replete with stories and ads full of such buzzwords as "neural marketing," "consumer brain scanning," and "targeted ads" and discuss the use of such diverse neuroscientific gear as eye trackers, EEGs, and even fMRI. If applied properly, such things can provide insight into purchasing decisions. (At least purchasing decisions made while covered in electrodes or strapped onto a gurney in a narrow tube with 100 dB sounds being made around your head for twenty minutes. A fluffy bunny jumper? Yes, I'll take two—*just get me out of here!*) But it's not all about marketing and sales. What if these targeted brain hacks were applied to create a calming sound field in a stressful environment, such as a hospital waiting room or an airport waiting area? Or if spatial sounds were applied to limit motion sickness on transportation? Or if sleep-inducing sounds were played for post-surgical hospital patients to help them recover faster? This is the beginning of the next age not only in understanding sensory and auditory processing in the brain but in making it useful in the real world. We have just begun to realize the practical potential of all the data collected in the thousands of neuroscience labs across the world. And just as sound has been with us from the start, it will help us define the future of our technology, and our lives.

Chapter 9

WEAPONS AND WEIRDNESS

SHORTLY AFTER LEAVING Columbia University in 1981 as an undiplomaed junior (and having been told I had no future in science by one of my biology professors), I was working as a gigging musician. I had just bought my first synthesizer, an old analog Juno 106, which, as was typical in the primeval techno-logical era of the 1980s, was no more digital than the cockroaches that infested my apartment. I was trying to figure out if it was possible to create fractal sounds—a largely theoretical construct at the time involving generating recursive tones that modified themselves in real time. Given the technology of the time, there was no way to change parameters on the fly using software—I was stuck just using the analog oscillators, filters, sliders, wheels, and other quaint manual controllers. The Juno was connected to a very old, rather noisy, but very, very big amplifier.

I had just spent an hour moving the controls to very precise positions to try to convince the oscillators to go in and out of sync while simultaneously shifting the low-frequency oscillator to act as a filter cutoff when my cat jumped on the control panel, rearranging things more to his liking. He then left at high speed, probably inspired by a very loud, very frustrated non-fractal vocalization on the part of his owner. Once I calmed down, and upon realizing that the cat probably had as much of a chance of coming up with fractal sounds as I did with this system, I pressed a single key. The amplifier gave out a very loud, weird sound, a low-frequency, high-powered beating thrum. It wasn't so much fractal as nauseating. I started to feel strange. So I turned to face the amp. And I threw up.

After cleaning up after myself, the amp, and the cat, all I could think was, "What the hell was *that*?" Somehow the cat and I accidentally made a sound that had a *very* non-auditory effect. Years later I was trying to assure a student that there was no such thing as the "brown sound," a legendary tone that was supposed to make you lose bowel control, when I remembered that scene. I started slowing down in my explanation about how sound was just something you heard and it couldn't affect your intestinal tract or your body, until finally, mumbling something about having to get to another class, I started wondering about the secret, sometimes classified aspects of sound.

Humans have a strange relationship with sound. We ignore most of it, shuffling it around in our brain stem to be monitored as "background," something that sets the stage of our world, driving attention but rarely capturing it. Yet even as we ignore it, it sits under our mind, doing all sorts of things that bubble up to our consciousness only after we put it in context of the rest of our senses and environment. It's the sensory system that runs in the dark, out of line of sight, telling you not so much what something is but that something important to your survival happened. Perhaps that's why humans often ascribe powers to certain types of sound, as if the waves themselves were alive and had intentions, good and bad.

Sound has long pervaded human cultures, linking us to the world. The Bible dictates that the world was created when God *said*, "Let there be light"; Hindus believe in the creative vibration of the syllable "om"; and the Huang Chung or "yellow bell" served as the primal tone of the scale that defined the relationship between the ancient Chinese imperial court and the harmony of the universe. This tendency runs under everything we do, from religion to pop culture, from science to military matters. And as with any tool humans pay enough attention to and play with long enough, these powers often get turned to dark uses. For every time sound is proposed as a healing device or as a way to induce social cohesion, there are probably two occasions when it gets used as a weapon, in both our stories and our reality.

The idea of sonic weapons is a very old one. In the hundreds of thousands of years P.I.,* our ancestors knew that sounds told them something was happening, usually something out of their control. The loud low rumbling of an earthquake or avalanche, the sudden snap of a twig and low growl of a predator in the dark, or the screaming of the wind during a hurricane all gave them warning of things powerful and to be avoided. And some ancestor of ours 17,000 years ago, long before sonic blasters were dreamed up for *Star Wars*, realized that if you took a flat piece of wood, put a hole in it, attached a long string to it and swung it over your head in a circle, it would create a low-pitched buzzing, roaring sound that changed as it got closer to or farther from the listener, a sound that people back then might have thought of as a helicopter had there been an accident with a time machine. This sonic weapon, the bullroarer, has been found in

---

* Pre-iPod.

archaeological sites on every continent but Antarctica, and in-digenous Australians still use it in rituals to frighten away evil spirits. While the technology is simple, the sound it makes is loud and complex, based on the low-frequency oscillations of the wooden airfoil spinning on the string, and shifting up and down in frequency due to the Doppler effect as it approaches and retreats from a listener on its circular path. The sound of a big bullroarer is very much like that of a large animal vocaliz-ing and breathing—hearing it at night around a fire certainly might have convinced listeners that some large, probably dan-gerous thing was invoked by this twirling sound source, and they could only hope that it was on their side against other things in the night. The bullroarer was thus one of the first uses of sound as a psychological weapon, even if it was aimed at some supernatural entity.

Sonic weapons have appeared in stories and real-world ap-plications in every part of the world, from the ancient bullroar-ers to contemporary non-lethal weapons such as the long-range acoustic deterrent (LRAD), which has been used (or misused) against modern-day crowds. Their various applications have relied on technology from bits of bone or wood on a string to high-tech phase arrays of piezoelectric emitters. But unlike many other weapons, the stories and fear behind sonic weapons are often more effective than the weapons themselves. Their cultural pervasiveness tells more about our hidden psychologi-cal dependence and relationship to sound than does their actual effectiveness in the world.

The first class of sonic weapons that people in the West prob-ably ever heard about were physical ones. According to a com-bination of biblical stories and archaeological research, around 1562 BC Joshua fought the Canaanites at the city of Jericho. The

battle was apparently quite fierce, ending with the walls of the
city crashing down. According to the biblical version, a divinely
inspired Joshua had his warriors simultaneously blow on their
shofars—rams' horns that can be played rather like a valveless
trumpet—followed by a shout given by all the warriors at the
same time. The massed common harmonic vibrations of the
voices are supposed to have collapsed the walls, allowing Joshua
and his people to do what touring bands always do: they destroyed the city completely, aside from Rahab the harlot, who
apparently helped set up the gig.*

This story is one of the earliest about the use of acoustic
weapons that could physically destroy something. It's easy to
understand why people might believe that a loud enough sound
could knock a building down. A team of horses thunders into a
village and pots get knocked off tables. An earthquake rumbles
and a building collapses. A crack of thunder is heard close by
and down the path a hut catches on fire. It's a short mental association from hearing a sound and experiencing some destruction to believing that the sound in the air, which we perceive
more easily than vibrations through the ground, is responsible
for the damage. But while sound is based on transmission of
energy by means of vibrating molecules, it takes a pretty specialized piece of equipment and a lot of power in a contained
space to get a substantial physical impact from airborne sound.

So would it have been possible for the massed vibration of
horns, voices, and silence, properly intermixed, to bring down
the stone walls of a city? The short answer, when you talk to architectural engineers, archaeologists, and acousticians, is no. The
rams' horns in question in the story of Joshua can put out a very

* Groupies have always had a special place in the hearts of bands on the road.

loud sound (about 92 dB) with a fundamental frequency of about 400 Hz, with most of the harmonic power between 1,200 and 1,800 Hz, according to acoustician David Lubman. This puts the shofar at about the power and annoyance level of a male bullfrog. A human male voice can run from about 60 Hz to about 5,000 Hz, with most of the energy below 1,000 Hz, and a maximum loudness of about 100 dB when screaming at the top of your lungs. This makes us louder than a bullfrog, but not by much. Lubman actually carried out an analysis of the acoustic scenario of Jericho, presuming that there were 300 to 600 men all playing shofars with superhuman precision and then all shouting in unison. He found that this would still not generate even a millionth of the power needed to induce structurally damaging vibrations even in the stiffest, most brittle materials, even if they were right in front of them, not to mention the exquisite timing required to get the phase of the sounds aligned enough to induce resonance in structures of different masses and sizes. But it does make a good story that tells us more about our relationship to loud sounds than it adds to the historical record.

But sound that propagates through denser materials is a different story. Living about a hundred yards from an Amtrak line, I can attest to the destructive power that 700 tons of train cars moving at 30–40 mph can transmit through the ground, and I have the cracks in my walls and ceilings to prove it. Sound traveling through a high-density medium moves much faster than it does in air, and so it attenuates more slowly. And if that vibration is in a closed structure, such as a metal-framed building, it's not uncommon to have the structure resonate—vibrate sympathetically with the initial source of sound. If it's allowed to continue, the resonance can build to the point that the structure fatigues and in theory could be damaged.

A story of this kind of destruction from modern times comes from the experiments of one of the most brilliant (and occasionally twisted) inventors of the modern era, Nicola Tesla. Born in 1856 in Smiljan, a village in what is now modern Croatia, Tesla emigrated to the United States and became known as the father of wireless transmission and alternating current. Among his fascinations was the idea of resonance, both electrical and acoustic. According to several biographies and innumerable conspiracy-theory websites, in 1898, Tesla was working with a small electromechanical oscillator that he attached to an iron pillar in his loft laboratory on East Houston Street in New York City. The story goes that as he let the oscillator run, the floors and space started resonating, causing small objects and furniture to start moving around the room. What happened next requires some skeptical inquiry: supposedly the resonance was propagating throughout the substrate of the building and then proceeded to start vibrating adjacent buildings and shops, blowing out windows and frightening people in nearby Little Italy and Chinatown; it did not stop until Tesla was alerted by the police pounding at his door. When asked if he had any idea what was causing the disruption, Tesla claimed (in one version) that he told them it was probably an earthquake, then showed them out and proceeded to destroy the oscillator with a sledgehammer, stopping the artificial earthquake.

When reporters arrived a bit later on, Tesla, with his typical tact and modesty, told them that he could have destroyed the Brooklyn Bridge in a few minutes if he'd wanted to. However, in a later article in the *New York World-Telegram* in 1935, Tesla gives a slightly different version of events, claiming it took place in "1897 or 1898." "Suddenly," he wrote, "all the heavy machinery in the place was flying around. I grabbed a hammer and broke the machine. The building would have been down about our ears in another few minutes. Outside in the street there was pandemonium. The police and ambulances arrived. I told my assistants to say nothing. We told the police it must have been an earthquake. That's all they ever knew about it." While Tesla was an inventive genius with more than 700 patents to his name (including patents 514,169 and 517,900 for the oscillating engines supposedly at the heart of this story), he also was known to be rather expansive in his claims about how simple his devices were and how terrifying their results, such as his ability to "bring down the Empire State Building with 5 pounds of steam pressure" and one of his oscillators.

So could this "earthquake machine" actually work? In theory, yes. Simple periodic vibration of rigid materials will cause them to resonate, and this resonance can get drastically out of control. Do yourself a favor and look up the video of the collapse of the Tacoma Narrows Bridge. The bridge was built in 1940 by Leon Moisseiff, and four months after it opened, a constant 42 mph wind induced aeroelastic flutter, a condition in which the natural resonance of a structure can enter a positive feedback loop if it is insufficiently damped in the presence of wind stress. The bridge began vibrating, the entire span tracing out elegant loops and waves at its natural resonance frequency until it collapsed into the river below.*

In a more modern example, the London Millennium Bridge started being called "the wobbly bridge" soon after its opening in 2000. At its ceremonial opening, the bridge began oscillating side to side as a resonance response to all the foot traffic and had to be closed to prevent a major structural disaster. So vibration

---

* The dog you see in the video survived the experience.

THE UNIVERSAL SENSE

can cause resonance disasters. But would Tesla's pocket-size oscillator have been able to cause an artificial earthquake in lower Manhattan? An attempt to simulate the experiment on *Mythbusters* in 2006 showed that a well-tuned linear actuator (an oscillating motor that runs back and forth as opposed to rotating) was capable of creating significant vibrations in a large metal bar, but a replica of Tesla's device had no major effect at any significant distance.

The *Mythbusters* team, never one to shrink from a challange, went full out and attached the oscillator to an old iron truss bridge, tuned it very precisely to the resonant frequency of the entire bridge, and were able to detect rumbling and vibrations quite some distance away, but the bridge itself was quite unaffected. They declared the story to be a myth. As with many such stories, the problem lay not with the underlying theory but rather in all the details of the real world. You could probably do major damage to a building or buildings with the proper vibration, but you need the power of a steady 42 mph wind or thousands of human feet stamping along, or at least an earthquake. Simple resonance may amplify vibrations in a rigid closed structure, such as a building frame, but every real-world structure is affected by damping, changes in vibrational ability because of changes in materials or environmental conditions. (This is what saves my house from falling down when the 2:00 A.M. freight train, which weighs considerably more than the Amtrak passenger train, goes by.) Even if Tesla had gotten the iron pillar in his lab to oscillate, that pillar was embedded in concrete somewhere, surrounded by soil, and separated from other buildings in the neighborhood. Even vibrations at the proper resonant frequency for an iron pillar would have been distorted and reduced once they hit lower-density materials, damping out to annoying rumbling sounds, which would be more likely to evoke noise complaints from the neighbors than to have entire neighborhoods running for their lives.

But given our mental history and neural wiring, we keep thinking that sound should be physically destructive on a large scale, and our more inventive engineers, scientists, and writers keep trying to come up with acoustic weapons. This is probably why they are such a mainstay in science fiction. In David Lynch's movie of Frank Herbert's *Dune*, the followers of the Kwisatz Haderach used a specialized "weirding module" that turned their voices into offensive weapons capable of blowing down walls, similar to Joshua's troops. In the film *Minority Report*, the police used handguns with "sonic bullets" capable of knocking people over. But my favorite fantasy sonic weapon was actually a toy built in the 1960s by Mattel—the Agent Zero M Sonic Blaster. In the tradition of 1960s toymakers' overkill in the name of fun, it used compressed air to create a deafening 150+ dB blast, capable of blowing over cardboard buildings and unraking huge piles of leaves with a single shot. And while it wasn't exactly a world-class military device, it definitely should be categorized as a sonic weapon given the number of cases of permanent hearing damage and perforated eardrums reported by the parents of the lucky kids who had them. This is all well and good if your goal is to deafen a few children and knock over a cereal box or two. But the ability of sound to destroy physical objects is fundamentally limited; it's in other realms—the psychological and physiological—that sonic weapons are real and effective.

Why create a sonic weapon? As a species, we've invested tremendous time and capital in creating weapons while simultaneously publicly decrying their use. Most weapons are based on the idea of using energy to convert organized solid structures into less organized and less solid structures. While acoustic weapons aren't that useful in this realm outside fiction, the effects that sounds have on us are useful for doing something analogous—taking an organized mind capable of plotting and carrying out plans, and using its own neural wiring against it. The basics of sound as a warning signal are built into us, from the deepest part of our brain stem to our highest cognitive centers. To destabilize a brain in a powerful manner, all you have to do is add on a few basic psychoacoustical features. Remember what evolutionary biology has taught us: if a sound is loud and low in pitch, it will be frightening. If it is loud and has lots of random frequencies and phase components (like fingernails on a blackboard), it will be annoying. If it is loud and inescapable, it will be confusing and disorienting. And if it is out of context, like a sound coming from somewhere it shouldn't, it is disorienting and frightening. The most effective acoustic psychological weapons combine all of these.

Think about almost every movie or cartoon you've ever seen where something is plunging down from the sky, whether it's a missile, a meteorite, or a hawk diving down on its hapless prey. There's almost always some sound effect to heighten the feeling of speed and descending danger. Yet in real life, aside from the extremely rare plummeting blobs of lava from a volcano, incoming asteroid, or, on the battlefield, incoming mortar fire (which is very new in auditory evolutionary terms), things that come down on you from above rarely make any noise that do any good as a warning. As a result, anything that makes a noise from above as it dives toward you is particularly frightening.

Imagine a scene in China in 220 AD. Soldiers in the Three Kingdoms under the command of Zhuge Liang hear a strange whistling sound and are suddenly pelted with flaming arrows. Chinese military inventors of the Three Kingdoms era invented the "whistling arrows," bow-launched arrows with a hollow head that made high-pitched screaming sounds as they flew through the sky. The earliest ones discovered were made of carved bone and actually had blunt heads, acting primarily as signaling and communication devices for troops, according to the annals of Sima Quian, prefect of the Grand Scribes during the Han dynasty. Later whistling arrowheads were more precisely made: they had smaller, sharper holes, were constructed of iron, were cast with single or multiple sharp points, and often were capable of holding flaming oil-soaked cotton. Their role had clearly changed from a long-distance communication and homing device to an acoustically charged attack weapon, so now they were not only screaming death from above but setting you on fire in the process. Their effectiveness in striking fear into those at the receiving end was expressed by Sun Guangxian sometime between 900 and 968: "A whistling arrow beyond the clouds, enough to shake the spirit even of the brave." As an acoustic terror weapon, they were enormously successful for over a thousand years, and archeological digs have found them scattered throughout China, Korea, and Japan.

In more modern times, the sound of danger from above was probably most notably used and remembered in World War II, with the introduction of the Nazi Junkers Ju87 bomber, more commonly known as the Stuka. As if a high-powered dive-bomber designed to strafe and bomb troops and civilians on the ground were not terrifying enough, the Stuka was fitted with a special device, a wailing siren called, appropriately, the Jericho Trumpet. The Jericho Trumpet was a siren activated by a small propeller that drove air into the sounding chamber, which created a kind of roaring sound that increased in loudness and pitch the faster the aircraft flew in its diving run, only cutting off when the airbrakes were applied and the craft leveled off. The psychological effect of Stukas diving down onto their targets, screaming louder and higher in pitch as they approached, only to go quiet after delivering their deadly load, was a major propaganda tool for the Nazis and became so ingrained in survivors and military personnel that it has become the iconic sound effect in a great many films showing any attacking or crashing aircraft.

The terror induced by the Stukas and the later V-1 "buzz bomb" didn't go unnoticed. In the post–World War II era, as the threat of nuclear war changed the face of direct military conflict, intelligence and military organizations started spending more attention and money on tactics that relied more heavily on manipulating enemy troops and civilians. So were born various psychological operations groups, more commonly known in the US as psyops. Psyops grew out of the Central Intelligence Agency's Special Activities Division (SAD), which carried out psychological and paramilitary activities, and by the 1960s almost all branches of the military had their own psyops groups. One of the more interesting programs was used in the Vietnam War; it highlights both the creativity of trying to use complex sound to manipulate a population under stress and many of the problems, most insurmountable, that such a technique must face. Traditional Vietnamese and other Asian people believe that if a person is not buried in the homeland, his or her soul will wander tormented forever, and "The Wandering Soul" was a recording that attempted to use this belief against the Vietnamese population.

The original tape, created by psyops engineers in 1968, was

about four minutes long and included traditional Vietnamese funeral music, spooky wailing voices of children calling for their fathers, and the voice of a woman talking about her husband being pointlessly killed in fighting for the North. Variants used recorded tiger sounds and different scripts, alternating between laughing and crying children and mournful sounds, all heavily modified with reverb and echoes to create a ghostly effect. The tapes were broadcast late at night from large multispeaker arrays out the side of helicopters and aircraft of the U.S. Army 6th Psyop Battalion, in the hopes that it would make villagers lay down their arms and guerilla fighters surrender rather than continue fighting. The mentality behind it was summed up as the unofficial motto of the 5th Special Operations Squadron: "Better to bend the mind than destroy the body."

But how effective was it? While to you reading this in the twenty-first century it might sound like a Halloween sound-effects CD, imagine hearing such a tape played very loudly, over and over, late at night in a jungle area after weeks or months of having people try to kill you. The psyops command structure apparently was heavily invested in it, believing that the combination of culturally specific sounds associated with death and loss and the latest electronic special effects would render the recording unearthly and terrifying. But they apparently missed one aspect: that the tape would play only when there was a clearly audible aircraft engine noise. Any human anywhere hearing ghostly sounds playing along with the sound of propellers and rotors is going to make the association pretty quickly that this is not a supernatural event and will rapidly shift their attention from the unearthly and unlikely sounds to respond to the military threat of an armed aircraft. In fact, eventually the typical response was that as soon as "The Wandering Soul" tape started playing, the local soldiers would just start targeting the acoustic source and bring most of their firepower into play trying to down it. This led to a change in military tactics in which "The Wandering Soul" was used as a lure to get ground forces to fire on it, followed by a second aircraft firing upon the now easy-to-locate ground forces. To put the final nail in the coffin of this acoustic weapon, the second gunship often played another recording called "The Laugh Box," which was a very irritating laugh track, after taking out the attacking ground forces. So while "The Wandering Soul" was an acoustic weapon, its role changed significantly as both sides grew accustomed to its presence.

But that was not the end of intelligence and military attempts at using sound as a weapon, always with mixed results. On December 20, 1989, Operation Just Cause began with the 82nd Airborne landing at Torrijos International Airport near Panama City, the goal being the capture and overthrow of Manuel Noriega. The operation itself was over in less than a week, but the complication was that Noriega had been granted sanctuary in the Vatican embassy. American psyops loudspeaker teams began broadcasting loud music, and according to newspaper articles at the time, the material broadcast varied from the sound of fingernails on a blackboard to high-powered American rock and roll at just sub-deafening levels (the first song apparently being "Welcome to the Jungle" by Guns N' Roses). The music played was supposedly driven by requests made to the Armed Forces Radio DJs and included high-powered presentations of songs such as "I Fought the Law" by The Clash, "If I Had a Rocket Launcher" by Bruce Cockburn, and "Hair of the Dog" by Nazareth (although on December 25 it supposedly was Christmas music played very loudly). This was labeled by *News-*

week as the "the most ridiculous psychological operation in U.S. history."

The problem is that while it sounds like someone was using basic loud noise and repetition to stress out Noriega and get him to leave his sanctuary, interviews with several military personnel who were there pointed out that the loud music was not there to manipulate Noriega. Rather, it was there to provide masking noise so that the dozens of press with their parabolic and other long-range microphones could not pick up sounds from meetings and report on negotiations that were being held inside to get Noriega to leave peacefully. The psyops speaker teams were still using loud music as an acoustic tactic, but it was more along the lines of a disinformation or security-enhancing technique than a method of directly affecting the players.

Some in the intelligence community saw opportunities for acoustic weapons that took a much more personal approach. In the 1970s, there were several interesting patents issued that described using modulated microwave energy to act as a carrier to beam sounds directly into a listener's head at a distance. The few direct sources that describe this technique describe it as a potential communication device that would prevent eavesdropping by any outside source, communicating directly from the source to the listener's brain without any external acoustic energy. However, given the proclivities of the intelligence community in this era, it's been suggested that it was more likely that this would be used as a psychological disrupter, making someone think they were hearing voices or sounds in an attempt to drive them mad.

Could such a system work? In a way, yes. You can perceive microwave energy, but not really in a useful manner. The first indication that people could hear electromagnetic energy arose during the early days of radar, when operators claimed that if they walked in front of radar emitters, they would hear distinctive clicking sounds. Subsequent studies by Ken Foster in 1974 showed that subjects actually could detect microwave energy. Unfortunately, it wasn't some odd neural receptor tuned to the electromagnetic spectrum. It's just that sufficiently powerful microwaves do what you think of microwave ovens doing—they heat different parts of the ear at different rates. So the listener is hearing the sound of his inner ear cooking.

But the idea that you could control the amount of thermal damage while modulating the microwave beam at an acoustically appropriate level lives on, not as a super-secret communication channel or a psyops device to make you think you're hearing voices, but rather as a purported crowd control device. The Sierra Nevada Corporation is planning on building what they describe as a non-lethal microwave ray gun, the Mob Excess Deterrent Using Silent Audio (MEDUSA). Initially developed with a military innovation grant, the idea is that the device could be used to deliver loud enough clicking sounds to drive people away from an area. The problem is that to create sounds that are even barely audible above background, you'd have to deliver about 40 watts of microwave energy per square centimeter. Considering that the safe limit for microwave exposure is about 1/1,000 watt per square centimeter, this sounds a lot more like a recipe for cooked humans than it does an effective means of crowd control.

There are much better ways to get sounds delivered in a tight radius at long distances, and that's by using piezoelectric speakers. Directional piezoelectric speakers work on a different principle than the diaphragm-based speakers we're more familiar with. Rather than using an electrical signal to induce changes in the strength of a magnet to vibrate a cone, piezoelectrics use special materials that vibrate directly in response to an electrical charge. A major advantage to them is that they can vibrate much faster than a mechanical speaker, allowing them to emit sounds far up in the ultrasonic range. Sounds with shorter wavelengths are less prone to diffraction, meaning they tend not to spread out as easily as sounds with longer wavelengths. So as long as you pour enough power into them, you can end up with a very long and relatively narrow acoustic "spotlight" (bearing in mind that the higher the frequency, the shorter the range at a given power level).

These piezo directional speakers typically use frequencies from 60 kHz to 200 kHz, far above the range of human hearing. In order to embed an audible signal into such a high-frequency carrier, you actually play two separate signals through the speakers, one a steady reference tone at the carrier frequency (say, 60 kHz) and a second one that is amplitude-modulated (AM) by an audible signal, varying between 60 kHz and 80 kHz. When the two signals strike your ear, they interfere with each other constructively and destructively, leaving only the difference between the two, at frequencies audible to your ear. You can experience this kind of ultrasonic modulation in many museums—these type of speakers let you walk in front of a painting or sculpture and hear a description of it that only projects in a very narrow area, using very high-frequency carriers to limit its travel so that it doesn't interfere with sound in the rest of the area and you don't get overlap from descriptions of other exhibits.

So could these be used as an acoustic psychological weapon? Maybe. Years ago I purchased one of the commercially available units, the Holosonics Audio Spotlight, with the idea of using it in field work. One of the issues you face in trying to study animal calling behavior is the difficulty of sending a signal to just one animal as opposed to an entire group. My idea was to use it to try to send signals to flying bats or calling frogs. Problems arose almost immediately. For the bats, 60 kHz is right in the middle of the echolocation range, so a half-meter-wide beam of 60 kHz sound was the bat equivalent of shining a strong spotlight right into your eyes. They all avoided it.* On the other hand, while the frogs didn't care about the carrier signal, the amplitude modulation technique used to get the audible signal mixed into the carrier doesn't really work on signals below 300 Hz, where much of the frogs' acoustic energy lies, so they just sat there and ignored it.

This left me with an expensive acoustic toy, time on my hands, and something of a mad-scientist bent, so I of course brought it down to my wife's loft in Williamsburg and hung it out the window, attached a microphone to the input, and waited for people we knew to walk down the street. The first time I tried it on a friend about halfway down the block, I used it to whisper that he owed me $50; his reaction was to immediately turn around and look for me, then to scan the street to see where the voice came from. I waited until he'd calmed down a bit and come closer, at which point I said, "Or was it $100?" His reaction was totally worth the money I never got back from him. He rapidly scanned the area again and then ran for the door and started pounding on it. When I let him in, I asked if he was okay, because he seemed a bit on edge. (I lose more friends that way.)

---

* Which did, however, make me the inventor of a long-range device to use on protesting bats.

And that's the problem with using these types of acoustic tools for psychological manipulation—my friend knew that anything weird that happened to his ears was likely my fault, especially if there was new equipment around. In the same way, people who have experienced this type of constrained sound—in museums, airports, casinos, and hotels—are not too likely to think they are hearing voices from a supernatural source or even inside their own heads; they are going to think that they are experiencing some technological effect. Several companies are even exploring tying video cameras with facial recognition software into use with directional speakers to play individually targeted ads to people walking by stores. The scenarios of science fiction movies are beginning to be played out, targeting not your sanity but your wallet.

Still, long-distance directional speakers have played a role in acoustic weapons technology—not for putting voices into people's heads but for using our own fearful reaction to loud, irritating sounds against us. Stories about acoustic weapons began emerging after 9/11, when almost every online and print news source began talking about unusual weaponry that played horrifying sounds, including those of fingernails on a blackboard or a baby's scream played backward at high levels, meant to drive anyone in its range away. While anyone who has been stuck in an airplane or other enclosed space while a baby screams knows how irritating this sound is, few people actually open the door in midflight to get away from it.

These types of stories kept cropping up for the next few years, less and less in legitimate news outlets and more and more on conspiracy web pages, until in 2004 when police on duty for the Republican National Convention in New York City were observed carrying a large flat black disk on a vehicle. This gadget matched descriptions of high-powered piezoelectric speakers that were just becoming commercially available. While the device was not used at the time, the Long Range Acoustic Device (LRAD) was first used in the United States in Pittsburgh in 2009, on protestors at the G20 summit. The LRAD is different from audio spotlights in that while it uses a phase array of piezoelectric speakers to create a tight cone of sound that can project for a long distance, it uses not ultrasonics but rather sounds around 2 kHz, just about in the center of humans' best auditory range. It creates an extremely powerful and piercing sound that is likely to activate just about every fight-or-flight circuit in the human brain.

People who have been close to the units describe an intense desire to get away, with some describing extreme nausea and panic, both responses modulated by elements of their sympathetic nervous system. Anyone within about 100 meters of an active LRAD is also in danger of suffering temporary and possibly permanent hearing damage. The LRAD has been credited by some as having a role in helping stave off attacks from pirates at sea; however, the distances between the two ships involved suggests that it was more useful as a warning signal than as a direct factor in getting the pirates to surrender. Unfortunately, its role has more often been to clear protestors away from an area; most recently it was used against Occupy Oakland protestors in 2011, with little regard for the long-term hearing damage or emotional trauma suffered by those subjected to deafening sounds directed against them. So acoustic weapons that work at a psychological level are here and in the hands of police and military units, but their long-term utility is yet to be determined. As Stephen Colbert pointed out, while these devices are useful against naive targets, what happens when people stop panicking in the face of a new device and start wearing noise-suppressing headphones to protests, or just putting their fingers in their ears? Will the acoustic arms race amp up into finding ways to actually injure targets without killing them all, allowing those weapons to wear the fundable label "non-lethal?"

Sound can certainly affect people at a physiological level, and not all of these effects are secondary biochemical ones, such as flooding people's brains and bodies with panic-induced biochemicals such as adrenaline and cortisol. The right kind of sound can actually cause physical damage to the ears and other parts of the body. Sound at 120 dB is at about the threshold for pain. Your ears are full of free nerve endings that act as pain receptors to warn you of potentially damaging events, ranging from noise above a certain level to the fact that a Q-tip is way too far in your ear canal. At 120 dB, the amplitude of the vibration of the air molecules is creating pressure changes extreme enough that your eardrum is stretching out of control and inner hair cells, particularly the high-frequency ones nearest the eardrum, are in danger of being ripped free of their moorings.

Your ear has a pressure relief system, called the round window, that flexes outward and inward as the fluids in the inner ear are pushed back and forth; however, even this system only has so much give to it. At sounds above this level, you start getting damage to the hair cells that causes your hearing to start losing sensitivity, starting at the high frequencies and slowly moving downward as the sounds get louder. If the duration of the loud sounds is relatively short or not too much above 120 dB, the loss of hearing is temporary and usually limited to certain frequencies. You can identify this damage by a loud hissing sound you'd hear for varying lengths of time, depending on how long you were exposed to a given sound (something I experienced after being too close to a few speakers at rock concerts). But if the sound goes above 160 dB, as when you're too close to an explosion or working in demolition or with heavy equipment without hearing protection, you get serious damage with permanent results, including ruptured eardrums, tinnitus (ringing in the ears), or even complete or partial permanent hearing loss. This is one reason why most consumer and professional amplifiers have a cutoff circuit, to make sure that you can't get sound above this level. And this is why the LRAD, which is supposed to be a psychological deterrent, can be put in the category of a physiologically damaging device: it puts out a whopping 167 dB at 1 meter, but because of the restricted nature of its beam shape, the dangerous level of sound extends for tens of meters in front of the device.

The type of sound that gets tagged as futuristically dangerous is usually ultrasound. I think there's something about the "ultra-" prefix that gets people cognitively and culturally excited and makes writers of science fiction feel a need to stick ultrasound weapons into futuristic scenarios ranging from the original *Star Trek*'s "Whom Gods Destroy" and "Way to Eden" through *Warehouse 13*'s supersonic zills in "Nevermore." It's probably helped along by ultrasound's association with high-tech medical imaging applications and numerous stories on conspiracy web sites about how ultrasound is used as a driver for psychological manipulation, as in the audio spotlight technology. But the uncool truth is that ultrasound has severe limitations as a weapon because of its limited range.

Ultrasound is defined as sound above the human upper auditory limit, around 20 kHz. While these high-frequency sounds are not as prone to diffraction as lower-frequency ones, their short wavelengths make them lose energy much faster. Getting ultrasound to work at any distance requires a lot of power. Bat echolocation uses ultrasonic frequencies up to 100 kHz blasted at 120 dB, giving it a range of about 30 feet. Some deep-sea dolphins use frequencies up to about 200 kHz emitted at about 130 dB, which might give them a range of 50–90 feet. But in both these cases, the ultrasonic signals are just being used to detect objects in the environment. For ultrasound to have any physical or physiological effect, it has to be very high-powered, very high-frequency, and very, very close. Under these conditions, ultrasound does pack a lot of energy into a very tight beam, not only capable of creating echoes for medical imaging but also capable of creating shock waves to blast apart kidney stones. This makes medical ultrasound sound like a small-scale sonic weapon. But the key is that medical ultrasound is using sound waves at millions of cycles per seconds, not thousands, and is close to the tissue being treated, helped by a gel that forms a liquid seal between the transducer head and the overlying skin. So what happens if you move the source further away? Could you use an ultrasonic blaster to blow something apart? I'm afraid not. Ultrasonic energy at a distance, even at very high power, will mostly just bounce off your skin. Even if you waved your blaster directly at someone's ear, it wouldn't even have enough power to move any of the hair cells, let alone rupture anything. So unless someone comes up with a power cell capable of emitting ultrasound at several million Hz and hundreds of decibels, you are safe from ultrasonic weapons.

But at the other end, in the infrasonic range, things are radically different. People don't usually think of infrasound as sound at all. You can hear very low-frequency sounds at levels above 88–100 dB down to a few cycles per second, but you can't get any tonal information out of it below about 20 Hz—it mostly just feels like beating pressure waves. And like any other sound, if presented at levels above 140 dB, it is going to cause pain. But the primary effects of infrasound are not on your ears but on the rest of your body.

There's an elevator in the Brown University Biomed building (hopefully fixed by now) that I've heard called "the elevator to hell," not because of destination but because there is a bent blade in the overhead fan. The elevator is typical of older models, a box 2 meters by 2 meters by 3 meters with requisite buzzing fluorescent, making it a perfect resonator for low-frequency sounds. As soon as the doors close, you don't really hear anything different, but you can feel your ears (and body, if you're not wearing a coat) pulsing about four times per second. Even going only two floors can leave you pretty nauseated. The fan isn't particularly powerful, but the damage to one of the blades just happens to change the air flow at a rate that is matched by the dimensions of the car. This is the basis of what is called vibroacoustic syndrome—the effect of infrasonic output not on your hearing but on the various fluid-filled parts of your body.

Because infrasound can affect people's whole bodies, it has been under serious investigation by military and research organizations since the 1950s, largely the Navy and NASA, to figure out the effects of low-frequency vibration on people stuck on large, noisy ships with huge throbbing motors or on top of rockets launching into space. As with seemingly any bit of military research, it is the subject of speculation and devious rumors. Among the most infamous developers of infrasonic weapons was a Russian-born French researcher named Vladimir Gavreau. According to popular media at the time (and far too many current under-fact-checked web pages), Gavreau started to investigate reports of nausea in his lab that supposedly disappeared once a ventilator fan was disabled. He then launched into a series of experiments on the effects of infrasound on human subjects, with results (as reported in the press) ranging from subjects needing to be saved in the nick of time from an infrasonic "envelope of death" that damaged their internal organs to people having their organs "converted to jelly" by exposure to an infrasonic whistle.

Supposedly Gavreau had patented these, à la Tesla, and they were the basis of secret government programs into infrasonic weapons. These would definitely qualify as acoustic weapons if you believe easily accessible web references. However, when I started digging deeper, I found that while Gavreau did exist and did do acoustic research, he had actually only written a few minor papers in the 1960s that describe human exposure to low-frequency (not infrasonic) sound, and none of the supposed patents existed. Subsequent and contemporary papers in infrasonic research that cite his work at all do so in the context of pointing out the problems of letting the press get hold of complex work. My personal theory is that the reason that his work survived even in the annals of conspiracy is that "Vladimir Gavreau" is just such a great moniker for a mad scientist that he had to be up to something.

Conspiracy theories aside, the characteristics of infrasound do lend it certain possibilities as a weapon. The low frequency of infrasonic sound and its corresponding long wavelength makes it much more capable of bending around or penetrating your body, creating an oscillating pressure system. Depending on the frequency, different parts of your body will resonate, which can have very unusual non-auditory effects. For example, one of the ones that occur at relatively safe sound levels (< 100 dB) occurs at 19 Hz. If you sit in front of a very good-quality subwoofer and play a 19 Hz sound (or have access to a sound programmer and get an audible sound to modulate at 19 Hz), try taking off your glasses or removing your contacts. Your eyes will twitch. If you turn up the volume so you start approaching 110 dB, you may even start seeing colored lights at the periphery of your vision or ghostly gray regions in the center. This is because 19 Hz is the resonant frequency of the human eyeball. The low-frequency pulsations start distorting the eyeball's shape and pushing on the retina, activating the rods and cones by pressure rather than light.* This non-auditory effect may be the basis of some supernatural folklore. In 1998, Tony Lawrence and Vic Tandy wrote a paper for the *Journal of the Society for Psychical Research* (not my usual fare) called "Ghosts in the Machine," in which they describe how they got to the root of stories of a "haunted" laboratory. People in the lab had described seeing "ghostly" gray shapes that disappeared when they turned to face them. Upon examining the area, it turned out that a fan was resonating the room at 18.98 Hz, almost exactly the resonant frequency of the human eyeball. When the fan was turned off, so did all stories of ghostly apparitions.

Almost any part of your body, based on its volume and makeup, will vibrate at specific frequencies with enough power. Human eyeballs are fluid-filled ovoids, lungs are gas-filled membranes, and the human abdomen contains a variety of liquid-, solid-, and gas-filled pockets. All of these structures have limits to how much they can stretch when subjected to force, so if you provide enough power behind a vibration, they will stretch and shrink in time with the low-frequency vibrations of

---

* You can get a similar visual display, called phosphenes, by rubbing your eyes in a dark room.

the air molecules around them. Since we don't hear infrasonic frequencies very well, we are often unaware of exactly how loud the sounds are. At 130 dB, the inner ear will start undergoing direct pressure distortions unrelated to normal hearing, which can affect your ability to understand speech. At about 150 dB, people start complaining about nausea and whole body vibrations, usually in the chest and abdomen.* By the time 166 dB is reached, people start noticing problems breathing, as the low-frequency pulses start impacting the lungs, reaching a critical point at about 177 dB, when infrasound from 0.5 to 8 Hz can actually drive sonically induced artificial respiration at an abnormal rhythm. In addition, vibrations through a substrate such as the ground can be passed throughout your body via your skeleton, which in turn can cause your whole body to vibrate at 4–8 Hz vertically and 1–2 Hz side to side. The effects of this type of whole-body vibration can cause many problems, ranging from bone and joint damage with short exposure to nausea and visual damage with chronic exposure. The commonality of infrasonic vibration, especially in the realm of heavy equipment operation, has led federal and international health and safety organizations to create guidelines to limit people's exposure to this type of infrasonic stimulus.

Since different body parts all do resonate and resonance can be highly destructive, could you build a practical infrasonic weapon by targeting a specific low-frequency resonance and thus not have to carry around a heavy amplifier or lock your victim in an elevator car? For example, imagine I am a mad scientist (a total stretch, I know) who wants to build a weapon

---

* This is probably about the level of distorted sound my amplifier put out in the event described in the introduction, hence the rather visceral effect.

using sound to make people's heads explode. Resonance frequencies of human skulls have been calculated as part of studies looking at bone conduction for certain types of hearing aid devices. A dry (i.e., removed from the body and on a table) human skull has prominent acoustic resonances at about 9 and 12 kHz, slightly lesser ones at 14 and 17 kHz, and even smaller ones at 32 and 38 kHz. These are convenient sounds because I won't have to lug around a really big emitter for low frequencies, and most of them are not ultrasonic, so I don't have to worry about smearing gel on the skull to get it to blow up. So how about if I just use a sonic emitter that puts out two peaks at the two highest resonance points, 9 and 12 kHz, at 140 dB and wait until your head explodes? Well, it'll be a while. In fact, it's not likely to do anything other than possibly make a nice dry skull shimmy on the desk a bit, and it will do nothing to a live head other than make it turn toward you to see where that irritating sound is coming from.

The problem is that while your skull may vibrate maximally at those frequencies, it is surrounded by soft wet muscular and connective tissue and filled with gloppy brains and blood that do *not* resonate at those frequencies and thus damp out the resonant vibration like a rug placed in front of your stereo speakers. In fact, when a living human head was substituted for a dry skull in the same study, the 12 kHz resonance peak was 70 dB lower, with the strongest resonance now at about 200 Hz, and even that was 30 dB lower than the highest resonance of the dry skull. You would probably have to use something on the order of a 240 dB source to get the head to resonate destructively, and at that point it would be much faster to just hit the person over the head with the emitter and be done with it. So while we still cannot use infrasound to defend ourselves against dangerous severed heads and have not found the "brown sound" that would allow us to embarrass our friends, infrasound can cause potentially dangerous effects on living bodies—as long as you have a very high-powered pneumatic displacement source or operate in a very contained environment for a long time.

Sorry to be a spoilsport about sonic weapons. I've always wanted to be able to wire up a couple of speakers in my basement lab and run around blowing holes in things and chasing away supervillains, but most sonic weapons are more hype than hyper. Devices such as the LRAD exist and make effective deterrents, but even these have pronounced limitations. A handheld sonic disruptor will have to wait for some major breakthroughs in power source and transducer technologies. But the uses of sound in the future probably hold more interesting promise than the ability to destroy things.

Chapter 10

FUTURE NOIZES

WELCOME TO THE future of sound! Artificially grown ears! Sonic disruptors! Implanted microsonar systems! Infrasonic tunnelers! Listening to the sounds of the sands blowing on Mars! Harmonic irritable bowel syndrome treatments!

I love futurism. It's always at least 50 percent wrong, but it's often inspiring and gives you something to think about when reading the latest blog entry about a cure for something that is right around the corner even though it actually only worked on four mutant mice in one lab that had questionable hygiene practices.

One reason I'm patient with these studies is that I have a lot of sympathy for anyone trying to make predictions about what comes next. The biggest problem with writing about things at the cutting edge of any field is that the edge keeps moving up behind you. Either your predictions are outdated by the time they get printed or they look ridiculous when they end up being flat-out wrong. While I was working on an early draft of this book, I recalled a 1998 Gordon Research Conference on neuroethology I went to as a postdoc. One of the sessions was on progress in auditory neural prostheses, also called cochlear implants. Cochlear implants are basically small microphones connected to amplifiers and filters that convert sound into electrical impulses that are fed directly into the auditory nerve to compensate for a non-functional inner ear. The first simple, single-channel model was developed by Dr. William House of the House Hearing Institute in Los Angeles in 1961. The early units were not very useful; they would let the deaf "hear," but the sound quality was so poor that it was basically seen as an adjunct to aid in lip reading. Thirty-seven years later, at the time of this conference, there were twenty-four-channel cochlear implants that could in some cases provide sound that was good enough that the profoundly deaf could actually hear and enjoy music as well as understand speech, and the devices continue to improve to this day.

The thing that started one of the louder and more interesting arguments was a point raised by Edwin Rubel of the University of Washington, one of the world's experts on noise-induced hearing damage and one of the pioneers of hair cell regeneration studies. After people had talked about the positive aspects of cochlear implants, Ed got up and gave a very heated talk about what a bad idea such implants were. His point was that once we have the technology and biopharmaceutical protocols for regrowing damaged hair cells, anyone who has gotten a cochlear implant will not be able to take advantage of such a treatment because the very act of inserting metal electrodes into the auditory nerve caused so much damage to the ear's infrastructure that there wouldn't be enough functional tissue left to try to regrow anything. He took the very strong position that if he had grandchildren who were deaf, he would strongly advise their parents *not* to let their kids get cochlear implants because we were "on the verge" of being able to regrow hair cells and presumably restore normal biological hearing to the deaf. The very vocal objections made after he said this were ones that come up often in debates about clinical practice versus research. Cochlear implants are here now, and the only animals that show the ability to naturally regenerate their ears are still fish, frogs, and birds. Over the years the clinicians seemed to be right—there were millions spent on research, but little progress on actually regrowing hair cells in adult mammals.

So when I started writing this chapter on the future of sound and hearing, one of the points I wanted to cover was the future of neural prostheses. Since 1998, neural prostheses have been gaining ground in reliability and complexity, in part due to greater understanding of the brain, but also in part due to better biocompatible materials, which cause less damage when left in the body for long periods of time. Neural prostheses are being used to prevent epileptic seizures and limit chronic pain, and some have even allowed locked-in patients to have limited interaction with the environment. They seemed to be the direction of the future (and still are). But then, an article that many of us thought we wouldn't read for 20 more years appeared in the journal *Cell* when Kazuo Oshima and colleagues at Stanford managed to grow functional hair cells from mouse stem cells in vitro.

Being able to regenerate hair cells is one of the most important future goals of auditory neuroscience. With 600,000 to 800,000 functionally deaf people in the United States and about 6–8 million with severe hearing impairment, being able to restore hearing is a major clinical goal. While cochlear implants are important and successful ways to do this, only about 250,000 of these 8 million are good candidates for the procedure, not to mention the fact that the surgeries cost about $60,000 per ear. The idea of being able to regrow sensory hair cells lost to disease; injury, developmental issues, chronic noise exposure, or just getting older is one that keeps hundreds of auditory and developmental neuroscientists trying to figure out exactly why every almost every other class of vertebrates is able to do this while mammals are not. The mammalian inner ear split away from the plan used by our other vertebrate relatives several hundred million years ago, and the adaptations that have given us a much wider range of hearing have come with a cost. The mammalian cochlea goes through a much more complex morphological maturation process than is seen in any other vertebrate, leaving less cell cycle-entry room to create new hair cells or to allow transdifferentiation of preexisting supporting cells into functioning hair cells. In addition, wherever hair cells have died due to injury or aging, the underlying tissues form scars that prevent the endolymph, the potassium-rich fluid of the inner ear, from leaking into cellular spaces where it could depolarize and damage other cells. But this scar formation also prevents any chance of regeneration of hair cells by natural processes. The loss of hair cells also leads to a loss of input to the sensory neurons in the spiral ganglion of the cochlea, which causes those neurons to retract and eventually die.

But if we can grow functional mouse hair cells in a petri dish, does this mean that in a few years we'll be able to transplant working human hair cells back into the cochlea and restore lost hearing? Well, no. In more than a few years? Maybe. First, the hair cells grown in culture were derived from mice, and mice, while mammals, have radically different functional and developmental pathways than humans. Mice mature in a few weeks and start showing signs of age-dependent hearing loss at anywhere from three months to a year depending on the strain. Humans with normal development and who have escaped noise-induced damage start showing age-dependent hair cell loss at about forty years of age. This means that since we don't normally regenerate them, our hair cells have some unknown mechanism for maintaining themselves at least forty times longer than happens in mice.

Several studies have suggested that these protective mechanisms might be exactly what prevents us from growing new replacement hair cells in the first place, and could easily prevent transplanted ones from taking. In addition, because of the tonotopic layout of the cochlea, we would not only have to grow human or human-compatible hair cells but surgically place them extremely precisely in the damaged area. The cochlea is one of the most complicated neural and sensory structures in the body, and the type of microsurgery required for such precise implantation in a living ear without damaging other parts of the ear has not yet been invented. Lastly, even if we replaced the hair cells, we would then have to replace the spiral ganglion neural connection from the hair cell to the cochlear nucleus via the inner ear. While some animals, like frogs, can regrow appropriate connections after damage to the auditory nerve, this is another thing that mammals are not too good at.

So if we're not likely to transplant actual hair cells, what about their progenitors, stem cells? The idea behind using stem cells is that they are pluripotent—in other words, an appropriately chosen stem cell could, in the proper biochemical environment, be transplanted into a damaged area and differentiate into the needed tissue. Researchers have been able to isolate cells from adult vestibular tissue (the balance part of the inner ear) and coax them into growing and differentiating into elements found in the inner ear such as neurons and glia, as well as showing some hair cell protein markers. Other studies have shown that by providing the proper culture conditions you can get embryonic (as opposed to somatic) stem cells to differentiate into hair-cell-like structures. But for either type of cell, the results from these studies have contributed more to the understanding of natural growth and differentiation than to the development of a functional therapy. The few studies that have tried actually transplanting stem cells into cochlear tissue have yielded little or no success. So despite jaw-dropping progress, the ability to biologically restore normal hearing is definitely part of future history.

So given that we're probably going to have to wait another few decades for stem cell transplants to become effective treatment techniques, what are some other possible directions for biological hearing restoration? If you read the "future directions" segments of scientific papers on the subject, you'll find projections based on current research, often giving you an accurate pathway for the next few years. These papers are usually written either by grad students who are putting together massive compilations of the most contemporary research in order to determine which way their own careers should go (while fulfilling publication requirements of their training labs) or by very senior researchers who have led the direction of a field for decades and are laying out the questions remaining to be addressed by current or future colleagues. In either case, they are limited by a few factors: the ideas have to directly build from current findings, they have to fit within current funding parameters, and they usually focus on things that the researchers or their immediate circle of colleagues are skilled at.

I've found one of the richest sources of futuristic ideas for hearing (or anything else) to be the informal gatherings, usually over drinks, after a conference, and informally run by undergrads, grad students, and postdocs (as long as the lab directors and funding agency representatives don't show up and spoil the fun). The last such auditory fest I attended after sneaking away from other members of my lab and buying a few rounds had some amazing suggestions for what comes next. One idea was to transplant an entire fetal anlage, the collection of cells destined to become the whole inner ear, directly into a damaged cochlea, where, when provided with intravenous culture medium, it would be able to grow an entirely new ear, using the damaged one as a template. When someone pointed out the ethical and political issues involved with harvesting an entire human embryonic structure, let alone some undifferentiated stem cells, another person suggested implanting not a human auditory anlage but one from another creature such as a frog. The argument here was that xenotransplantation—the transplanting of organs from another species—has been going on for more than a century, ranging from the early twentieth-century transplant of goat testicles into a male human scrotum to help "low male urges" through the accepted contemporary practice of using pig heart valves as replacements for damaged human ones. This surgery would almost certainly be rejected by the human host's immune system, but still—it was a great example of innovative thinking.

Another suggestion was to use temporarily implanted microinjectors to introduce promoters that would reactivate some of the identified genes underlying normal cochlear structural development as well as cell death promoters. By judiciously applying one at one end and the other at the other end, you could theoretically grow a new cochlea while simultaneously digesting the old one. The problem with this approach was that while many genes have been identified that help lay down structural axes and the way in which body parts grow out of them in a regular fashion, actually triggering them in a controlled way to grow a functional body part is still far down the road. My contribution was to examine mammals that live abnormally long lives, such as bats and naked mole rats (both of which live three to five times longer than they should by any current metabolic models) and see if they have any protective mechanisms that enable them to keep hearing longer.

Bats in particular would be an interesting model because they are much more closely related to us than are mice, and because they are absolutely dependent on hearing to survive—a deaf bat will starve to death if it doesn't kill itself by flying into a tree at the wrong moment. In addition, bats' hearing is all high-end (as far as we know). Understanding how they preserve their 20 kHz hearing at thirty-five years of age might at least give us some insights into the nature of cochlear protection. None of these ideas are things you are likely to see written about as a scientific success story in the next few years, but these kind-of-out-there ideas are the inspiration for the next generation of auditory neuroscientists, hopefully even after they sober up.

One neuroengineering grad student was adamant about the fact that biological experiments *always* take longer to perfect than technological ones, and we should be focusing on technological adjuncts to our ears rather than trying to improve on 300 million years of mammalian evolution. There are indeed advantages to working with technology rather than biology, not the least of which is if you mess up an electronics experiment, you don't have to stanch bleeding.

Technological applications in the field of miniature and bio-logically compatible electronics have undergone staggering progress in the last ten years, often leading people who were on the cutting edge five years ago to wonder what happened. For example, shortly after finishing my Ph.D., I was invited to attend a Defense Advanced Research Projects Agency (DARPA) conference on acoustic microsensors. The driving idea behind it was a battlefield intelligence application to identify things in an area based just on their sound and vibration and upload the information to a remote server. The proposed system would use semi-independent acoustic modules that were small enough that hundreds could be dumped out of a low-flying aircraft and, upon hitting the ground, be able to form a network that could report on acoustic events. The acoustic basis for the modules were newly created Knowles subminiature microphones, a few millimeters on a side, capable of picking up a wide range of sounds. Bioacousticians who studied animal hearing were called in because animals, including ourselves, are excellent at identifying sounds and figuring out where they are coming from. Based on the projected network parameters, remote listeners would be able to identify the type of sounds using frequency and amplitude analysis and know the location of a sound based on differences in amplitude and phase between the microphones.

The hope was that if you spread enough of these around an area, you could pick out individual events, such as the low-frequency rumble of approaching soldiers. The idea was a fascinating one, but it suffered from a couple of problems. One was that in 1998 there weren't readily available batteries of the right specs or tiny low-power networkable broadcast devices, but these were "just engineering issues," as scientists like to say when they aren't the ones who have to solve the problem. The bigger issue was in the choice of invitees to the conference. The attending bioacoustic scientists were among the cream of the crop for cutting-edge animal hearing science, but most of the animals they studied were mammals—in other words, animals that predominantly hear high frequencies. Mice, chinchillas, bats, and cats, species who were the research focus of most of the attending scientists, are great at detecting and locating airborne sounds, but not so good at identifying vibrations that travel through the ground. What was needed were experts in frogs, scorpions, and naked mole rats, animals that are much more sensitive to low-frequency ground-based sounds.

That project, like many DARPA projects, didn't end up coming up with a practical solution to the question at hand, but it did provide many of the participating labs with access to amazing technology. Direct and indirect spin-offs from the program eventually yielded things such as subminiature microphone arrays capable of localizing sounds a significant distance away to within a meter or so of accuracy (which was unheard of at the time), a plethora of automatic sound categorization algorithms, and the recent development of a biomimetic cochlea-like chip that picks up radio frequencies in a manner similar to that of the mammalian ear. The miniaturization of microphones continues on, and the 2 mm microphone of 1998 is now down to a barely visible 0.7 mm, with equally reduced power requirements. Specialized audio transducers such as ultrasonic sonar emitters and detectors and underwater hydrophones have likewise dropped in size and price.

So what can you do with these tiny sound devices? The market for miniature microphones for use in personal electronics alone is staggering. Almost 700 million were sold in 2010 for use in everything from cell phones to personal computers, military communication devices to industrial robotics. And as reliability has increased and the price has dropped, they show up in more and more applications. One application demonstrated as a proof of concept in 2002 was the idea of an implantable cell phone. The idea was that a small radio-frequency chip powered by a miniature battery could use a subminiature microphone/speaker implanted at the tooth/bone interface to both send signals to the inner ear and pick up spoken words via a bone conduction pathway through the jaw. You probably would not want to let your teenager get one of these when they become available (what are you going to do, take his jawbone away when he goes over his minutes?), but embedding such a device into a removable cap or bridge could be extremely useful for coordination and communication between members of search-and-rescue teams, police, or military, not to mention the people being searched for or protestors on the other side. At some level, the embedded cell phone is the culmination of a few decades of miniaturizing consumer electronics, which we've been seeing since the advent of the Walkman.

Another piece of acoustic technology that has benefited from miniaturization is the ultrasonic transducer, the basis of sonar. Even just ten years ago, most ultrasonic transducers, capable of emitting and picking up signals above 20 kHz, were either very simple single-frequency devices such as the inch-wide Polaroid units used to focus cameras or very expensive and delicate research-grade devices used in underwater sonar applications. But in the last five years, it's become easy to find devices much smaller and more powerful than these research units in almost any electronics or robotics store, most attached to inexpensive amplifier and detector circuits that will let a homemade robotic toy avoid objects just a few inches across up to 10 feet away. We've started seeing small sonar devices in cars to keep you from running into the back of your garage, sonar "tape measures,"* and even wearable miniature sonar platforms that can be built into clothing or hats to aid the visually handicapped. The miniaturization of ultrasonic transducers has also allowed them to be used for better non-invasive medical imaging of ever smaller structures and even engineering applications to allow detection of flow through tiny pipes and tubing to detect leaks. It's not too far a stretch to imagine that within a few decades (or much sooner) wearable high-definition sonar units will be able to fit into headbands to allow people to carry out search-and-rescue operations in the dark, or that surgeons will be able to wear micro-miniature sonar units on their fingertips or deploy them on scalpel heads to be able to generate three-dimensional views of the surgical area projected onto a heads-up display to make surgery safer. And coupled with advances in neural prostheses, it likely will not be too much further down the road before we will be able to take sonar data and convert it into signals that the human visual cortex can understand, giving us a bat- or dolphin-like ability to see in the dark and into each other's bodies.

But focusing too much on the use of sound in highly technical fields ignores the huge role sound plays in the everyday lives of the billions of people with daily access to smartphones and the Internet. During a discussion about the *Just Listen* project, Brad Lisle and I were talking about how sound not only could be used in teaching and media applications but also be an interactive tool to get people more engaged in their sonic world.

---

* The downloadable sonar ruler iPhone app uses lower-frequency sounds giving a resolution of only a few inches—when it works.

One of the projects we got most excited about was one we called Worldwide Ear. Worldwide Ear is a proposed citizen science project to carry out acoustic environmental mapping using simple recording equipment that will provide publicly available crowd-sourced information on global bioacoustics. The project would be run in a similar fashion to other web-based citizen science experiments such as Galaxy Zoo (www.galaxyzoo.org), which allows users to identify galactic phenotypes in Hubble images, or Snowtweets (www.snowtweets.org), which allows users to tweet the depth of snow in their area and then observe global snow depths based on the collected data using a application called Snowbird. Anybody with a cell phone or recording device and Internet access could participate, starting with entering information about the make and model of their recording device and their location. Two to three times a day, a user would go to a specific location, orient and mount their equipment in the same way, and record sounds for a minimum of five minutes. The recording would then be uploaded to a central server, where it is translated into a common format (e.g., MP3) and linked to a geographical database program with user-uploadable content, such as Google Earth or NASA's World Wind.

Why would this be useful? Why would people want to bother participating in such a project? Because by getting enough people to participate, we could create a remarkable environmental tool to assess global acoustic ecology. Acoustic ecology is the study of changes in sound caused by modifications to the environment, by either human or natural causes. Sound can be both a measure of environmental factors, such as the loss of specific birdsongs in an area that has undergone significant development, and an instigator of environmental changes, as shown by correlations between poorer cardiovascular health and excessive noise levels in inner-city areas. As changes in sound are among the most pervasive and ubiquitous things we can sense, the Worldwide Ear project would allows us to examine the ecological "health" of a region over time scales ranging from hours to years with relatively simple equipment.

If the Worldwide Ear project gets under way, it would not only give us an aural window on our world, letting us be acoustic tourists, but also could be an incredibly powerful research, policy, and educational tool. Such a freely accessible acoustic database could provide lawmakers and acousticians with information on urban bioacoustics, the sonic environment of cities. It could let someone compare low-frequency noise bands at different times of day in Rome, a populated site with a great deal of vehicular traffic, versus Venice, a site with many fewer roads but comparable human foot traffic. These data would be important for assessing human health and epidemiology by allowing researchers to create maps of specific urban and rural regions based on sound levels for specific frequency bands (similar to isobars used in meteorology to create weather maps) and then look for correlations with human health or cognitive issues. Do quiet areas have lower cardiac risk? Do students in schools near airports show lower performance?

The data would also be useful for understanding economics. Are farming communities that use "bird cannons" showing better crop yield by frightening off birds or are they showing lower yields because of increased insect pest activity from the lack of birds? Can correlations be made between acoustic conditions and socioeconomic status of a region, that is, are property values higher in quieter areas, or is there more wealth in regions with noisy manufacturing? And what about monitoring the ecological health of an area? Recordings made at different times of the day from different sites can be used in combination with automatic animal call recognition software to identify the species of birds, frogs, insects, and other acoustically active animals and map changes in population activity across a year or even changes in population density by comparing the sounds of a single species over multiple years or locations. While this is a near-future application that could be implemented today, the data acquired could help researchers assess and improve human and ecological health on Earth for decades to come, just by people listening where they live with the tools they already have.

But what can go beyond mapping the whole Earth with our ears? Hearing is our exploratory sense, the one that reaches out ahead and behind us, in dark and light. So it is somewhat ironic that its role has been so incredibly limited in the greatest human adventure of all, the exploration of space.

At the ungodly hour of 6:30 A.M. on October 9, 2009, my wife and I headed over to the NASA Planetary Data Center at Brown University to watch the LCROSS spacecraft impact into the lunar surface in aid of NASA's search for water on the moon. Because of the hour, I was a little concerned that no one would show up at the event. I was very happy to see that I was wrong—two rooms were full of students and faculty, eyes locked on the screens showing the real-time link from the LCROSS spacecraft on NASA TV. When the final countdown started, the room became silent. And there, in the middle of a black region untouched by sunlight probably for billions of years, was the tiniest of blips on the screen, so small that I wondered if it was just signal interference. The room was full of quiet murmurs, mostly from people who missed it, wondering if something had gone wrong.

I realized that one of the things making many of us wonder was that this huge impact yielded only the tiniest bit of input to any of our senses. While intellectually everyone in that room *knew* that there wouldn't have been any sound even if they had been standing on the moon within sight of the impact (except for vibrations transmitted through their space suits' boots), the silence divorced us from the event.

We expect sound when something huge happens—and if there is none to be had, we provide it with applause and cheers (or screams and yells). For every extraterrestrial landing event, the success of the landing, out of sight millions of miles away from the Earth, with no access to sound or images, is usually signaled by a flip of a data bit, but heralded by the cheers of the mission planners. (Or the silence, which weighs heavier and heavier when the signal doesn't come through, as with the Mars Polar Lander.) For anyone old enough to remember it (and allowed to stay up late enough to watch), the first moon landing was a pivotal event. For a brief moment, no one cared about the Vietnam War, student protests, or racial problems. Everyone was watching a human set foot on another world. But what stands out for me, and what I still carry burned into my memory, is not the blurry video signals but the noisy, static-ridden, and highly compressed voice of Neil Armstrong saying, "That's one small step for a man, one giant leap for mankind." (And yes, he did say the "a"—subsequent analysis of the old radio signals a few years back showed that it was there but had gotten lost in data compression.)

But these are the sounds humans provide. The entire arena of sound off Earth is something that has gotten only the most cursory attention, probably because we think of space as silent. Sound requires some medium to propagate, and human ears are primarily sensitive to airborne sound. But while interplanetary and intergalactic space have vacuum levels so dramatic that we have trouble simulating them even with expensive test chambers, they are not truly empty. There are particles out there, a few hydrogen atoms in each cubic meter—not much compared to the 10^{25} particles in the same volume at sea level on Earth, but with a powerful enough source, it's still possible to get what would pass for an acoustic oscillation. The problem is that propagation of acoustic waves in space happens over such large distances, over such long time scales, and in such a thin medium that to hear the B-flat drone of a black hole fifty-seven octaves below middle C (giving it a period of oscillation of over 10 million years), you need either really big ears and a lot of patience or access to the NASA Chandra X-ray observatory, as Andrew Fabian, the drone's discoverer, had.

For the most part, when you hear what's called "sounds from space," most of what you are actually hearing is "sonification," or the translation of electromagnetic wave phenomena such as radio waves into acoustic signals. You need the translation for a number of reasons. First of all, you don't have any sensors capable of picking up electromagnetic radiation.[*] Next, while radio information has frequency, period, and amplitude and is subject to a lot of the same loss factors as sound, including reflection, refraction, and spreading loss, radio signals are spreading and vibrating about a million times faster than sound does on Earth.

But we have over a century of experience converting radio signals into sound right here on Earth, and nearly eighty years

***

[*] It's not biologically impossible—electric fish use electric fields to communicate via specialized electroreceptive sensors on their skin.

ago Karl Jansky built the first operational radio telescope, allowing him to listen in to the radio emissions of the Milky Way. Since then, sonification of electromagnetic and space-borne acoustic events has let us listen in to the sounds of our sun and other stars ringing as convection currents cycle heat from their surface, and listen to the howling of the solar wind as charged particles spread out through the solar system. Ground- and satellite-based systems let us hear the effects on the Earth's upper atmosphere of this solar wind, generating eerie high-pitched kilometric radiation from auroras, as well as the "Earth chorus" formed by free electrons spiraling through the Van Allen radiation belts. Large radio telescopes and orbital probes have detected the same kind of phenomenon around Jupiter and Saturn, not only giving us insight into the structure of their electromagnetic environment but showing commonalities of planetary interactions with the solar wind in both the inner and outer parts of the solar system. And as we've pushed the limits of our remote exploration further and further, sonification by John Cramer of the University of Washington of the microwave energy left over from the big bang has let us "hear" the 14-billion-year-old echoes of the creation of the universe, a slowly changing, mournful sound, as if the universe were having second thoughts.

But despite our endless fascination with the energies of interplanetary and interstellar space, we are planet dwellers. Even I have to admit to being much more interested in what goes on or under the surface of Mars or in the ethane lakes of Titan than in the incredible power of a solar coronal mass ejection, even though the latter might impact me more directly by knocking out my GPS. Our acoustic experience on the planets has been extremely limited. Despite eighteen successful landing missions on other bodies in our solar system, only three probes have had dedicated microphones built in.

In 1981, the Soviet Union launched the Venera 13 and 14 probes to take measurements of the atmosphere and carry out experiments on the surface of Venus. Previous Venera missions had landed and survived on the surface for up to two hours, but all of them had their share of problems, ranging from stuck camera lens covers to failure of the soil analysis experiments due to damaged pressure seals. But Venera 13 in particular was remarkable in that it not only survived the crushing atmospheric pressure and high temperatures but sent back high-resolution images of the surface, analyses of soil samples drilled from the ancient basaltic ground, and, for the first time, sound from another world. The Venera 13 and 14 probes had an instrument called the Groza-2, designed by investigator Leonid Ksanfomaliti, which consisted of seismometers for detecting surface vibration and small microphones for picking up airborne sound. The microphones were heavily armored and relatively insensitive, designed more for survival in the pressure-cooker atmosphere than for high-fidelity recordings, but they worked for the several hours of descent and several more on the surface, in the midst of sulfuric acid clouds. The microphones detected the sound of thunder as the probes were descending, and the low susurration of slow, thick winds on the surface.

I tried for several years to get copies of the actual recordings, but neither the original tapes nor any backups seem to have been translated into any contemporary format. The closest I could come to hearing the original sounds were low-sample-rate plots of the waveforms showing how the Groza-2 instrument on Venera 13 picked up the sound of the lens cap ejecting and striking the surface, followed by the sound of drilling during a soil sampling experiment and the sound of the sample being placed in the experimental chamber. Before we roll our eyes at such a limited return, bear in mind that the recordings were done using technology from thirty years ago on the surface of Venus, at temperatures of about 455°C and under 89 Earth atmospheres of pressure. I have trouble getting decent recordings when it's raining.

The only other solar system body to reveal any of its sounds was not even a planet but Titan, the largest moon of Saturn. Titan is a peculiar moon, 50 percent larger than our moon, making it almost planet-size, tidally locked with its host planet, Saturn, giving it a day of 15 days and 22 hours in Earth time. Its distance from the Sun, 1.5 billion kilometers, would make it just another icy or rocky body, except for one thing: it has an atmosphere similar in composition to that of Earth, mostly nitrogen with traces of organics such as ethane and methane, which form its clouds. Rather than the thin, wispy envelope you might expect from a small body, its atmosphere is actually about 50 percent thicker than that of Earth. Under the influence of both energy from the sun and tidal stresses from Saturn, Titan is an extremely dynamic place, covered in lakes and rivers of ethane and propane, snows of methane, sand made of frozen hydrocarbons forming massive dunes, and slow-moving but powerful weather.

On December 25, 2004, a small cylindrical body separated from the American nuclear-powered Saturn probe Cassini and began its approach to Titan. This probe, called Huygens after the seventeenth-century Dutch astronomer Christiaan Huygens, landed in a "muddy" area on the surface of Titan near the region called Xanadu on January 14, 2005. The probe was battery-powered and the best that was hoped for was good data return during the two and a half hours it would take to descend through the thick atmosphere, and perhaps a few minutes of surface operation. During its descent, one of the most novel devices was the Huygens Atmospheric Structure Instrument (HASI), which contained accelerometers and a small microphone to capture the forces on the spacecraft during descent and the actual sounds of the Titanian winds. When combined with data from the Doppler Wind Experiment, which used radio telescopes to calculate the changes in the probe's position as it swung under its parachutes while descending, researchers were able to reconstruct the sound of the winds on Titan, giving us the first auditory glimpse of this far-off world. When I listened to them on the European Space Agency website, I got a very strange feeling when I realized that even more than a billion kilometers away, some things could still sound like home.

Aside from Veneras 13 and 14 and Huygens, only the Mars Polar Lander was equipped with sound recording equipment—an instrument called the Mars Microphone. This was a 1.8-ounce digital sampling device that was meant to record short sound samples on the surface of the red planet. With only 512K of onboard memory, it could hold a total of one 10-second sound clip, but for 1998, it was a masterpiece of audio miniaturization, not to mention radiation-proofing. Recording on Mars is a serious challenge compared to recordings on Earth, Venus, or even Titan, all of which have substantial atmospheres. The Martian environment, while almost Earth-like compared to the crushing pressure and melting temperatures of Venus or the frigid −180°C solvent-laden mud of Titan, has an atmosphere that is only 1 percent as thick as that of the Earth, so sounds are much quieter. The Mars Microphone had built-in amplifiers that would have boosted the signal up to audible levels, but, sad to say, it met its end when the Mars Polar Lander crashed on the surface due to a programming error. Its replacement was due to launch on the French Mars Netlander, but the project was cancelled in 2004 and now it simply sits idle. Given the low priority most researchers place on hearing the sounds of other places in the solar system, in part due to equipment and launch costs but also because of the relatively high bandwidth required to get real-time sonic information, it may be that the first people who actually hear the sounds of Mars are the ones who will land there. But, frankly, this is shortsighted. Because human space explorers evolved in the terrestrial acoustic environment, when they find themselves on an alien world, sound may not be so much a sensory warning and information source as a source of confusion.

Relatively little attention has been paid to the human psychoacoustics of space exploration, which is odd considering how acoustically stressful it can be. Even Yuri Gagarin, on the first human spaceflight, commented on the sounds of launch as an "ever-growing din . . . no louder than one would expect to hear in a jet plane, but it had a great range of musical tones and timbres that no composer could hope to score." Early space flight research here and in the Soviet Union did spend a lot of time and resources examining the effects of vibration and force on bodies during simulated launch stress, which is where some of the best work on infrasonic effects on human bodies emerged. Using ground-based data, NASA (and presumably the Soviets) carried out extensive analyses to determine the engineering procedures for establishing acceptable noise levels, effects of reverberation in enclosed spaces, and what kind of bandwidth communication equipment would need in order to avoid having critical verbal commands or alarms be masked by any spacecraft cabin noise. But there has been relatively little recent published work in the United States on noise and sound levels in spacecraft. Since construction of the International Space Station (ISS) is finally complete and it has full-size working crews on a permanent basis, trying to see if there are long-term effects on hearing and downrange cognitive effects would seem to be a useful endeavor. The ISS is significantly different from any previous spacecraft or habitats. With 837 cubic meters of internal space, it is the largest space structure built to date, and has been continuously occupied for over eleven years.

Despite constant harping about cost overruns and criticisms that it has been scientifically underutilized, mostly because of delays in completing it, the ISS has provided us with an amazing laboratory to study the effects of long-duration space habitation on humans. But one of the most overlooked stressors in any environment is chronic noise exposure. The ISS suffers from some of the same issues as any other spacecraft: it is basically an air-filled rigid container with fans that have to be active more or less constantly. This creates a constant resonance condition that can be only partially lessened by sound insulation. A paper by R. I. Bogatova and colleagues in 2009 reported that an onboard acoustic survey showed that noise levels were above the safety limits in every region of the crew module, from workstations to sleeping quarters. It may seem a minor point to worry about whether your spacecraft is too loud when you have to put up with 100 dB noise in subway tunnels or kids playing their music at deafening levels, but remember: this is the only place these astronauts can go, so the noise never stops. You can't exactly open the door and go outside (at least not without a lot of preparation, anyway), nor can you safely wear earplugs all the time because they might prevent you from hearing a critical alarm.

Yet acoustic stress can have serious long-term effects on task performance, emotional state, attention, and problem solving, no matter whether you're on Earth or in orbit. This is a particularly important factor to consider for the future, because human-crewed spacecraft that will one day go to Mars or beyond will probably be built much more along the lines of the ISS rather than the tight-fitting capsules of Soyuz and Apollo. We have to consider the role of acoustic stress in the daily life and abilities of crews when they start exploring beyond Earth orbit.

And what will they hear when they get there? I doubt that I will be around when the first humans hover in the Jovian cloud decks or slog through the frozen mud of Titan, but I still have hope that there will be human boots on Mars in my lifetime. To date, all extra-vehicular activities on missions have taken place either in Earth or lunar orbit or on the moon, environments with nothing to carry sounds at normal levels of human sensitivity unless the astronauts put their helmet against a structure. The heavily insulated nature of their boots would prevent anything but the strongest ground-based shock and vibration from getting through to them. But Mars will be different. While it is geologically (and biologically, we presume) much less dynamic than Earth, it is a planet with remarkable weather and a chemically active environment, and despite fifty years of attempts to decipher its secrets, it is still mostly unexplored. Gullies have been seen on canyon rims, showing evidence of possible flows of briny water, and ice caps of frozen carbon dioxide expand and retreat with the seasons, leaving strange landforms in their wake. In its more temperate regions, giant sand dunes change on a daily basis, marching around and through craters and valleys, while sub-surface tunnels lead into areas that may hold thicker atmospheres and more water. So Mars is not a dead place—it is merely very different, and so, presumably, are its acoustics.

As I've noted, the Martian atmosphere is only 1 percent as thick as Earth's, and it is mostly composed of carbon dioxide, with temperatures at the height of a human head ranging from 1°C to −107°C. These factors yield three basic differences from sound on Earth. First, the speed of sound would be about 244 meters per second, about 71 percent of the speed of sound at sea level on Earth. The second aspect is that the Martian atmosphere would tend to attenuate sounds in the range of 500–1,500 Hz, right in the middle of humans' most sensitive auditory region. Lastly, the lower density of the atmosphere would drop the relative loudness of any sounds by anywhere from 50 to 70 dB. We can presume that future space suits built for operations on Mars likely would have built-in microphones to monitor outside sounds. So our Martian explorers will have to learn to compensate for the acoustic differences in their new environment.

Even presuming that the microphones are connected to amplifiers, properly calibrated to prevent the masking of verbal communications from the ship or other team members, the explorers would still have to cope with the fact that everything around them will sound different. Imagine that they are standing on the surface near a crater wall when the vibrations of a rover trigger a rock slide. But where was the rover? Where was the rock? They will have trouble determining not only how far away a sound is but also exactly where it is coming from. We evolved our auditory localization ability based on differences in the time of arrival and relative loudness of sounds in our two ears based on the propagation qualities of sound moving through air on Earth; we can't even do it underwater on our home planet. The low speed of sound and the attenuation of frequencies in our best hearing range will cause problems on Mars. Even identification of the sources of sound will be made more difficult, not only because it will be a completely new environment full of strange sound sources but also because of the loss of spectral information.

Luckily, it won't interfere with voice communication (anyone dumb enough to remove his helmet and yell on Mars is probably going to have more to worry about than his voice sounding funny), but environmental sounds will not be easy to figure out. While the explorers are trying to get surveying equipment set up, a low-pitched moaning sound comes into their headsets from the outside. It's only seconds later that they feel a gentle pattering of sand on their helmets. Even the sound of the wind, normally almost white noise, will sound different, the notch in the 500–1,500 Hz range making it sound lower in pitch, more organic. Few astronauts are likely to assume it's the sound of some ravenous alien, but it will still be disorienting. Human space explorers will take our hundreds of millions of years of auditory evolution and hundreds of thousands of years of human psychophysics with them, and like their ancestors moving into any new environment, they will have to learn to listen for the sounds of danger or the sounds of opportunity. No matter where we go or how far in the future, one of our oldest, most conserved, and universal sensory systems will both adapt to and drive the evolution of how the human mind will cope with future scenarios.

Chapter 11

YOU ARE WHAT YOU HEAR

I'VE SPENT THE last thirty years of my life listening to things. Actually, check that: I've been listening to things for about half a century. I've only been *paying attention* while listening to things for the last thirty years or so. And over the last eighteen years, I've been paying a lot of attention to brains (human and otherwise), the sounds those brains are trying to assess, and the sounds they make.

I was lucky enough to get my training in things both musical and scientific at the right time. Despite years of piano lessons as a kid, I seriously got into music only in my early twenties, when the first PCs became affordable and useful and the idea of being able to connect them to musical instruments became practical with the advent of the musical instrument digital interface (MIDI). My entry into auditory science in the 1990s not only trained me in traditional psychophysical and anatomical techniques but also let me enter the field just at the point when EEGs changed from giant, clumsy, relay-driven, steampunk-looking towers to sleeker self-contained units, when PET and MRI scans were becoming practical tools for neural imaging, and when electrophysiology moved from the analog oscilloscope era to the more self-contained digital format. It let me play at the crossroads of sound technology, from making sounds to figuring out how they affect the brain. And I still remember my thought the time I carried out my first successful brain recording.

The brain sings.

During an electrophysiology experiment, the first thing that grabs my attention is the sound the brain makes. Is it a stream of white noise, meaning the electrode is not in yet? Or does it click in time to a stimulus, meaning I've hit an auditory responsive area? Are there huge tympanic strikes like a snare drum solo, independent of any sound I play to it, suggesting I'm in some region that has an important rhythm of its own but doesn't play well with outside sounds? Or are there soft, susurrating changes like the sound of waves on a beach, suggesting that maybe I pushed too far and the electrode is in a ventricle, letting me indirectly hear cardiovascular rhythms in the cerebrospinal fluid?

For over a century, placing a conductive electrode through a surgical opening in the skull into the brain has been *the* way to gather information about the real-time ionic processes that underlie neurons, which spend their days sending signals outward and receiving information sent inward. Most often the collected data are published as images: plots of amplitude of changes in voltage over time, audiograms to illustrate neural sensitivity to sounds at different frequencies, graphs of changing conductance of individual patches of channels, or diagrams showing connectivity between different responding areas based on the differences in timing of responses from two electrodes in different places. But more often than not, our first data are sonic, as the neuronal electrical changes are passed into a small stereo amplifier and to a perfectly normal speaker.

These data rarely make it into publication—I'm sorry to report that we don't yet have research journals that embrace multimedia enough to reproduce audio recordings. Admittedly, it's a very particular complaint, but I've always thought this is a shame. I can usually tell where I'm recording in the brain depending on how the brain itself sings. Depending on the type of electrode you use, you can listen in on millions of neurons at a time or the ticking of individual ionic channels. You can even identify different types of neurons by the sounds they make in response to a stimulus. If you're recording using an electrode with high impedance to pick out responses from a single neuron, and you play a brief clicking sound even to an anesthetized subject, you'll discover that the brain responds, issuing electrochemically based clicks of its own.* Some neurons will click back only at the start of the sound, some only at the end. Some will create a burst of regular clicks. Some will do nothing. Sometimes what you hear is not a response but only the highly distinctive "neuronal death song," which sounds like a plaintive wail (although it is really just ionic channels dumping potassium through inappropriate holes in the cell membrane from misplaced electrodes).

Each neuron has a characteristic sound depending on its condition and input/output state. But contrary to the increasingly passé "neuron theory," which suggests that if we could

---

* Impedance is the opposition to a time-varying electric current, so it is similar to resistance. The higher the impedance of an electrode, the smaller the volume you are recording from, and hence the fewer neurons you will be able to record from.

just get down to the smallest individual element of the brain we could finally see how it all works, those of us who spend a lot of time recording neurons realize that, individually, they don't mean much to a living organism. Outside the experimental recording area is a whole living brain with billions of neurons firing, each with its own electrical pattern based on what it and its neighbors and all the other neurons that seem to be unrelated to the stimulus are doing. Thousands of neurons die every day and few are replaced, yet cognitive or functional degradation takes decades to notice or, for some lucky portion of the population, doesn't change at all until the final off switch is thrown.

Complex computation, found in any brain more complicated than a few neurons floating in a culture dish, requires ensemble or population coding, hundreds or thousands or millions of neurons working in gangs to isolate features and characteristics of fuzzy input, figure out what bits go together, and put together complex perceptions on the way to something more interesting, such as consciousness. To encode (and decode) reality, which is a very noisy place, requires a system capable of taking the noise and changing it into useful signals with neuronal equivalents of filters and amplifiers, modules that while running small programs gather this noise and interact in loops of input, turning stochastic noise into quantal percepts—this tone, those smells, that color.

One mistake that people make when thinking about filters is falling prey to the idea that since filters remove information, the filtered output is necessarily less complex and more constrained than the original input. This underlies a lot of philosophical statements that the human experience is but a subset of reality—the path from physics to sensation to perception gets narrower as you go. But our perceptual filters no more act in isolation than do our neurons. The output from two (or two thousand, which is more biologically likely) filters, due to the multitude of integrations that occur from divergent neuronal populations, can and will interact. The filtered outputs may be parallel, in which case they will sum and strengthen specific inputs, or they may be completely opposite, generating neuronal and perceptual cancellation, but often they will resonate and beat against each other to create new response patterns. These neuronal filters are not blinders—they are what allow us to take the noise and get a handle on it at biological speeds, not worrying overly much about femtosecond changes in thermal energy of vibrating atoms and focusing instead on the lush harmonies of rising strings.

Depending on the brain's state (which depends on whether you're awake, asleep, aroused, thinking, reading this book, or scratching), certain neuronal activities will be amplified and others attenuated; however, some trends will remain across all conditions as long as the organism is alive. These differences are analogous to filters and amplifiers in a sound system and are based on individual differences in the brain. Your regular habits become encoded in your brain as defaults: you are comfortable paying attention to certain inputs and ignoring others. These defaults are flexible, allowing you to enjoy your favorite songs but be delighted at hearing a new one that you like or switching off one that makes your head hurt.

The differences between individual brains derive from development, environment, health, culture—almost anything that has made its mark on the living being. With 10 billion to 100 billion neurons to work with and trillions of synapses in play, strengthening, weakening, beating, and looping at biologically relevant rates, the outputs create a personalized neuronal signature, a sort of neuronal timbre, as individualized as the sound from a century-old musical instrument. If you dare to think of the brain as a whole, composed of billions of neurons, each with its own dynamic sound changing along with its function, the function of the brain as a whole is reflected in a metasound—a neuronal orchestra.

Think back to psychoacoustics and the chapter on music. A song is an epiphenomenon, a whole that is more than the summed aspects of the physical acoustics of the instruments, the architectural acoustics of the performance space and/or the recording gear, the level of skill of the musicians, and the effects of what they had for breakfast the morning of their performance. In the same way, all the individual firings of neurons, transportation of fluids, and activation and deactivation of genes that occur in the normal function of the brain give rise to something more than a collection of sonifiable neural responses—they produce a mind.

No one person, lab, or field understands the brain or the mind. But with 30,000 neuroscientists attending the Society for Neuroscience's annual conference, you get a feel for how incredibly deeply we want to understand it. Wandering through posters covering cutting-edge work before it's even published, hearing talks about new directions based on unexpected breakthroughs, realizing that entire careers can be focused on trying to comprehend how a single molecular channel microns across contributes to a structure that may have trillions of similar molecules, you begin to get some idea both of how much there is to learn and of how long it will be before we can truly put it all together. The workings of the human brain and the mind that arises from it are as much unexplored territory as interstellar space is.

There is certain uneasiness among a lot of neuroscientists about using the word "mind." While it is frequently thrown around, often somewhat sloppily, in cognitive or psychological papers, you don't usually hear a neuroscientist talking about the mind until well after she's gotten her Nobel Prize and it's safe to do so. As a non-Nobel winner, I tend to focus on the function of the brain, most likely the home of the mind. But the brain is no more the mind than the seed is the sunflower. It is the place from which the mind grows, develops, emerges, functions, and eventually fades. I and most of my colleagues search for the mind in our data—sheaves of electrical tracings, graphs of neural responses, ornate illustrations of stimulus and response. Those with a more molecular tilt search for its underpinnings in the slightest change in cellular behavior. Those who do neuroimaging hope that by viewing a living, unanesthetized working brain, they will glimpse that subtle epiphenomenon emerging.

But all of our studies tend to be problematic. The mind can't be found in the slow temporal smears of fMRI or the vast baroque operations of genetics. These are too far from the temporal domain in which our minds seem to operate. Nor has it been found in the fast-moving but spatially gross architecture of an EEG, or in the accurate but constrained single-neuron pulse of an electrophysiological recording. Over the last fifty years we have looked into working brains of humans and our kin; seen complex metabolic and electrical changes in response to subtle sensory and cognitive inputs; explored the biochemical underpinnings of thought, imagination, and language; crowed when a gene was found that seemed to underlie human specialness and then grumbled quietly when it was found in hundreds of other living things, doing subtly different things to and in their brains. And yet we are still identifying the pieces.

We are splitting the underpinnings of the mind into smaller and smaller elements, much as early particle physicists used larger and larger particle colliders to find smaller and smaller elements of matter. But the overview, a theory of everymind, still eludes us. And despite the technology we throw at it, in our recursive way, we still need to simplify things to try to figure out how to put all the disparate pieces together.

The one thing we can do with all these unassimilated pieces is build models. Models are small-scale representations of what we think might be happening, based not only on past research but also on our prediction of where our data will lead us. But our ability to predict the future is uneven, and models are fragile things. Breakthroughs can completely unravel years of scientific theories, so even bold models from brilliant minds are subject to constant revamping and reevaluation. This is the basis of honest science. There have been dozens of models proposed to describe how the mind works and what it is, or at least what it is *like*, and each of them arose from the public interpretation of what was cutting-edge science at the time.

During the age of telephony, the mind was described as a switchboard, bidirectionally connecting different brain modules the way an old telephone network would connect phones. In the 1950s, in the early days of laser experimentation and holography, the stability of the mind and the ability to retain memory even after traumatic brain damage led some to describe it as a holographic device, with every bit of the brain containing a holistic memory of the entire mind. In the 1970s, as computers began to become more commonplace in the lab, the mind was described as a computer, and shortly thereafter computers were designed to try to model aspects of how the mind worked, bringing about the first steps into developing artificial intelligence. By the late 1990s, as our ability to work more easily with genetics expanded with the introduction of polymerase chain reaction techniques and gene sequencing equipment, the mind started being viewed as something that emerged from the confluence of genetic tendencies and environmental conditions, a sort of neurogenetic entity. As we entered the twenty-first century and our computers, phones, and even books started merging into a single interconnected web of information, people started suggesting that intelligence and mind emerge from the sheer complication and mass of data processing. And now, as we enter the second decade of the twenty-first century, our wealth of information has brought us no closer to a really useful model of the mind. Some even suggest that the current focus on fMRI and some other forms of neuroimaging has sent us backward, substituting pretty pictures of living brains for an understanding of what those brains are actually doing.

I've thought about this for many years and had many discussions about it (and a few knock-down, drag-out fights). I'm convinced that if we really want to get to the root of what mind is, we should stop diving deeper into data and actually *think* about the knowledge that we already have, preferably in new ways. One of the advantages of trying to write a book that covers a subject as huge as sound and the mind is that you are forced to step out of your normal daily perspective and try to figure out how things work in the real world rather than just the lab. And during one of the days I spent thinking rather than writing, I remembered the minor war that broke out in my world when it was suggested that bats may be able to respond to sounds that differ only on the nanosecond scale (orders of magnitude faster than their nervous system could possibly cope with). This could have just been another one of those in-house battles between neuroscientists that never escapes the bounds of academia. But still, the problem bothered me: How can a bat respond to a change in a signal only a few hundred nanoseconds long when their nerves fire on a scale thousands of times slower than that, and by all the classical models of brain function can't even detect changes hundreds of times slower?

When I started thinking about that in a larger format, I realized something about brains in general: if we want to look for the mind, we've been looking in the wrong place and at the wrong time. We tend to scale our thinking about how the brain functions based on what we think of as its basic units—its nerves and, to a lesser extent, the supporting tissue associated with them, such as glia and blood vessels. But each nerve, depending on where it is in the brain, can have thousands or millions of predecessors, all contributing not just the yes/no of an impulse but slight modulations of the tendency to fire or not. What we detect with any of our narrow-sighted techniques is just a view through a tiny temporal window of an extraordinarily complex neurochemical orchestral arrangement, not just shifting the charged ions that flow in response to the order to fire or not, but creating states of potential change. Will the three hundred excitatory inputs to this hippocampal cell be overridden by the forty inhibitory inputs? Will the built-in gap junctions contribute enough to the currents to change an onset response to a series of periodic pulses that swamp further inputs? And will all this actually take that sound that I just heard and turn it into a memory or just put it on hold to compare it to what comes next or what has come before? At any given site in the cortex, where 10 billion neurons may have fired before, it may be that tension, that moment between when all the summed inputs have arrived and when an output is chosen—to sing, to say a word, to smile, to have that neurochemical rush that is the moment of epiphany—in which the mind hides. Think of it as a kind of mental quantum foam, where the possible probabilistic states of what will happen next collapse.

Our minds are built from our past experiences, whether biochemical or environmental. Our experiences as we develop cause us to amplify some types of experiences, wonderful or traumatic, and filter out others than have been of limited usefulness or interest in the past, modulated in intensity by choice and circumstances. In short, our minds are built moment by moment within the tension between input and output, formed by our life experiences just as the breath from a sax player shifts the energy in the body of the instrument, adding the reverberations of the metal and the room, amplifications and filtering of lungs, lips, face, and instrument, guided by hands. And what emerges is not just modulated air but music.

One of the best ways to discover new things is to look at them a new way. So: is the mind itself music?

The idea of the musical mind is not a new one. You can find the phrase in the titles and subtitles of at least fifty books and journal articles from professional journals as lofty as *Science*. But it usually refers to how behavior or neural function changes when exposed to elements of music or how the musically trained may be different from the musically naive. What I'm suggesting is that we think about the mind in the ways we think about music, as a process to be explored instead of a thing to be identified. The failure (so far) of science to define music comes from its focus on the infrastructure rather than the process and flow. A melody or song is not just a string of notes of known duration and loudness. It is not what activates the left versus the right amygdala. It is not what makes you smarter or dumber. It is the tensions between the discrete events and the flow of what comes next. And perhaps that is the mind as well, with the brain driving the loops of massed neuronal activities, each contributing a bit of filtering, a bit of amplification, a change of tempo to signal something worth attending to or an increase in noisiness when things get confusing. Perhaps *between* the moments of tension and release of neuronal activation and inhibition is the emergent portal of consciousness. The mind, like music, seems likely to me to be more about the flow of information than the information itself, popping into consciousness after all input has arrived, from the quantal to the conversational, when all input is processed, filtered, amplified, and streamed by the parameters that make up the emergent mind; the moment between the wordless thought and the word.

If we could record all the sounds of all the neurons in any given brain, every brain would play a song composed of moments of event-driven and cognition-derived sounds. What better way to think of this ever-changing yet highly personalized waveform than as the mind of the listener? This would be an incredible technological challenge. Practically speaking, you can't get 100 billion physical electrodes into someone's brain without converting it from the most complex information-processing system on the planet into a conductive pincushion. But given the rate of progress in biomedical recording and imaging, and scientists' and engineers' abilities to play with existing technologies to get even more powerful toys, perhaps it's not too far off. Conceptually? It's probably not *the* answer to the question of the mind, but perhaps it's a useful way to think of new ways of thinking about brain function and the mind.

If we could convert the song of the brain into something we could listen to, could we gain an intuitive sense of what is working or not working in an individual's mind? We do a simple version of this in the outside world already, letting us listen to the squealing cries of high-energy proton storms on Jupiter as a lonely spacecraft orbits, capturing the changes in its magnetic fields and replaying them for us to interpret as sounds. Would localized brain damage from a stroke sound like a song played with a section of the orchestra missing? Would we be able to hear early-onset Alzheimer's creeping up like a slow detuning of the strings in an orchestra? Would some forms of mental illness sound like harmonic distortion? Would a flash of insight sound like the rising chorus from Beethoven's *Eroica* or the prosodic tune of a voice yelling, "Eureka"?

I wish I knew. I hope I will one day as I keep studying sound.

But it's something to think about while waiting for the next song to come up on your playlist.

SELECTED SOURCES AND SUGGESTED READINGS

Chapter 1: IN THE BEGINNING WAS THE BOOM

Fritzsch, B., Beisel, K. W., Pauley, S., and Soukup, G. "Molecular evolution of the vertebrate mechanosensory cell and ear." *International Journal of Developmental Biology.* 51 (2007): 663–678.

Lewis, J. S. *Rain of iron and ice: The very real threat of comet and asteroid bombardment.* New York: Perseus Publishing, 1995.

Pieribone, V., Gruber, D. F., and Nasar, S. *Aglow in the Dark.* Cambridge, MA: Belknap Press, 2006.

Pradel, A., Langer, M., Maisey, J. G., Geffard-Kuriyama, D., Cloetens, P., Janvier, P., and Tafforeau, P. "Skull and brain of a 300-million-year-old chimaeroid fish revealed by synchrotron holotomography." *Proceedings of the National Academy of Sciences USA.* 106 (2009): 5224–5228.

Wilson, B., Batty, R. S., and Dill, L. M. "Pacific and Atlantic herring produce burst pulse sounds." *Proceedings of Biological Sciences.* 271 supp. 3 (2004): S95–S97.

Chapter 2: SPACES AND PLACES

Charles M. Salter Associates. *Acoustics: architecture, engineering, the environment.* San Francisco: William Stout Publishers, 1998.

China Blue. "What is the sound of the Eiffel Tower?" *Acoustics Today.* 5 (2009): 31–38.

Declercq, N. F., and Dekeyser, C. S. A. "Acoustic diffraction effects at the Hellenistic amphitheater of Epidaurus: Seat rows responsible for the marvelous acoustics." *Journal of the Acoustical Society of America.* 121 (2007): 2011–2022.

Chapter 3: LISTENERS OF THE LOW END

Boatright-Horowitz, S. S., and Simmons, A. "Transient 'deafness' accompanies auditory development during metamorphosis from tadpole to frog." *Proceedings of the National Academy of Sciences USA.* 94 (1997): 14877–14882.

Boatright-Horowitz, S. L., Boatright-Horowitz, S. S., and Simmons, A. M. "Patterns of vocal interactions in a bullfrog (*Rana catesbeiana*) chorus. Preferential responding to far neighbors." *Ethology.* 106 (2000): 701–712.

Mann, D. A., Higgs, D. M., Tavolga, W. N., Souza, M. J., and Popper, A. N. "Ultrasound detection by clupeiform fishes." *Journal of the Acoustical Society of America.* 109 (2001): 3048–54.

Simmons, A. M., Costa, L. M., and Gerstein, H. B. "Lateral line-mediated rheotaxis behavior in tadpoles of the African clawed frog (*Xenopus laevis*)." *Journal of Comparative Physiology A: Neuroethology, Sensory, Neural, and Behavioral Physiology.* 190 (2004): 747–758.

Tobias, M. L., and Kelley, D. B. "Vocalizations by a sexually dimorphic isolated larynx: peripheral constraints on behavioral expression." *Journal of Neuroscience.* 7 (1987): 3191–3197.

Weiss, B. A., Stuart, B. H., and Strother, W. F. "Auditory sensitivity in the *Rana catesbeiana* tadpole." *Journal of Herpetology.* 7 (1973): 211–214.

Yamaguchi, A., and Kelley, D. B. "Generating sexually differentiated vocal patterns: Laryngeal nerve and EMG recordings from vocalizing male and female African Clawed Frogs (*Xenopus laevis*)." *Journal of Neuroscience.* 20 (2000): 1559–1567.

Chapter 4: THE HIGH-FREQUENCY CLUB

Dear, S. P., Simmons, J. A., and Fritz, J. "A possible neuronal basis for representation of acoustic scenes in auditory cortex of the big brown bat." *Nature.* 364 (1993): 620–632.

Hiryu, S., Bates, M. E., Simmons, J. A., and Riquimaroux, H. "FM echolocating bats shift frequencies to avoid broadcast-echo ambiguity in clutter." *Proceedings of the National Academy of Sciences USA.* 107 (2010): 7048–7053.

Griffin, D. R. *Listening in the dark: The acoustic orientation of bats and men.* New York: Dover Press, 1974.

Horowitz, S. S., Stamper, S. A., and Simmons, J. A. "Neuronal connexin expression in the cochlear nucleus of big brown bats." *Brain Research.* 1197 (2008): 76–84.

Chapter 5: WHAT LIES BELOW

Aeschlimann, M., Knebel, J. F., Murray, M. M., and Clarke, S. "Emotional pre-eminence of human vocalizations." *Brain Topography.* 20 (2008): 239–248.

Davis, M., Gendelman, D. S., Tischler, M. D., and Gendelman, P. M. "A primary acoustic startle circuit: Lesion and stimulation studies." *Journal of Neuroscience.* 2 (1982): 791–805.

Emberson, L. L., Lupyan, G., Goldstein, M. H., and Spivey, M. J. "Overheard cell-phone conversations: When less speech is more distracting." *Psychological Science.* 21 (2010): 1383–1388.

Halpern, D. L., Blake, R., and Hillenbrand, J. "Psychoacoustics of a chilling sound." *Perception and Psychophysics.* 39 (1986): 77–80.

Hart, J. Jr., Crone, N. E., Lesser, R. P., Sieracki, J., Miglioretti, D. L., Hall, C, Sherman, D., and Gordon, B. "Temporal dynamics of verbal object comprehension." *Proceedings of the National Academy of Sciences USA.* 95 (1998): 6498–6503.

King, L. E., Douglas-Hamilton, I., and Vollrath, F. "African elephants run from the sound of disturbed bees." *Current Biology.* 17 (2007): R832–R833.

LeDoux, J. *The emotional brain: The mysterious underpinnings of emotional life.* New York: Simon & Schuster, 1996.

McDermott, J., and Hauser, M. "Are consonant intervals music to their ears? Spontaneous acoustic preferences in a nonhuman primate." *Cognition.* 94 (2004): B11–B21.

Olds, J., and Milner, P. "Positive reinforcement produced by electrical stimulation of septal area and other regions of rat brain." *Journal of Comparative Physiological Psychology.* 47 (1954): 419–427.

Paré, D., and Collins, D. R. "Neuronal correlates of fear in the lateral amygdala: Multiple extracellular recordings in conscious cats." *Journal of Neuroscience.* 20 (2000): 2701–2710.

Pressnitzer, D., Sayles, M., Micheyl, C., and Winter, I. M. "Perceptual organization of sound begins in the auditory periphery." *Current Biology.* 18 (2008): 1124–1128.

Ross, D., Choi, J., and Purves, D. "Musical intervals in speech." *Proceedings of the National Academy of Sciences USA.* 104 (2007): 9852–9857.

Shamma, S. A., and Micheyl, C. "Behind the scenes of auditory perception." *Current Opinion in Neurobiology.* 20 (2010): 361–366.

Chapter 6: TEN DOLLARS TO THE
FIRST PERSON WHO CAN DEFINE
"MUSIC" (AND GET A MUSICIAN,
A PSYCHOLOGIST,
A COMPOSER, A NEUROSCIENTIST,
AND SOMEONE LISTENING TO AN
iPOD TO AGREE . . .)

Bangerter, A., and Heath, C. "The Mozart effect: Tracking the evolution of a scientific legend." *British Journal of Social Psychology.* 43 (2004): 605–623.

Chabris, C.F. "Prelude or requiem for the Mozart effect?" *Nature.* 400 (1999): 826–827.

Fritz, T., Jentschke, S., Gosselin, N., Sammler, D., Peretz, I., Turner, R., Friederici, A. D., and Koelsch, S. "Universal recognition of three basic emotions in music." *Current Biology.* 19 (2009): 573–576.

Grape, C., Sandgren, M., Hansson, L. O., Ericson, M., and Theorell, T. "Does singing promote well-being?: An empirical study of professional and amateur singers during a singing lesson." *Integrative Physiological and Behavioral Sciences.* 38 (2003): 65–74.

Harris, C. S., Bradley, R. J., and Titus, S. K. "A comparison of the effects of hard rock and easy listening on the frequency of observed inappropriate behaviors: Control of environmental antecedents in a large public area." *Journal of Music Therapy.* 29 (1992): 6–17.

Malmberg, C. F. "The perception of consonance and dissonance." *Psychological Monographs.* 25 (1918): 93–133.

Plomp, R., and Levelt, J. M. "Tonal consonance and critical bandwidth." *Journal of the Acoustical Society of America.* 38 (1965): 548–560.

Rauscher, F. H., Shaw, G. L., and Ky, K. N. "Music and spatial task performance." *Nature.* 365 (1993): 611.

Seashore, C. E. *Psychology of music.* New York: Dover Publications, 1967.

Ventura, T., Gomes, M. C., and Carreira, T. "Cortisol and anxiety response to a relaxing intervention on pregnant women awaiting amniocentesis." *Psychoneuroendocrinology.* 37 (2012): 148–156.

Zattore, R. H. "Music and the brain." *Annals of the New York Academy of Sciences.* 999 (2003): 4–14.

Chapter 7: STICKY EARS

"The Use of Sound Effects." *The Radio Times—BBC Yearbook,* 194. London: British Broadcasting Corporation, 1931.

Szameitat, D. P., Kreifelts, B., Alter, K., Szameitat, A. J., Sterr, A., Grodd, W., and Wildgruber, D. "It is not always tickling: Distinct cerebral responses during perception of different laughter types." *Neuroimage.* 53 (2010): 1264–1271.

Meyer, M., Baunmann, S., Wildgruber, D., and Alter, K. "How the brain laughs: Comparative evidence from behavioral, electrophysiological and neuroimaging studies in human and monkey." *Behavioural Brain Research.* 182 (2007): 245–260.

Chapter 8: HACKING YOUR BRAIN THROUGH YOUR EARS

Graybiel, A., and Knepton, J. "Sopite syndrome: A sometimes sole manifestation of motion sickness." *Aviation Space and Environmental Medicine.* 47 (1976): 873–882.

Gueguen, N., Jacob, C., Le Guellec, H., Morineau, T. and Loure, M. "Sound level of environmental music and drinking behavior: A field experiment with beer drinkers." *Alcoholism: Clinical and Experimental Research.* 32 (2008): 1795–1798.

"Hell's Bells." *Maxim.* 100th issue, 166. 2006.

Karino, S., Yumoto, M., Itoh, K., Uno, A., Yamakawa, K., Sekimoto, S., and Kaga, K. "Neuromagnetic responses to binaural beat in human cerebral cortex." *Journal of Neurophysiology.* 96 (2006): 1927–1938.

Klimesch, W. "EEG alpha and theta oscillations reflect cognitive and memory performance: A review and analysis." *Brain Research: Brain Research Reviews.* 29 (1999): 169–195.

Lawson, B. D., and Mead, A. M. "The Sopite syndrome revisited: Drowsiness and mood changes during real and apparent motion." *Acta Astronautica.* 43 (1998): 181–192.

Rockloff, M. J., Signal, T., and Dyer, V. "Full of sound and fury, signifying something: The impact of autonomic arousal on EGM gambling." *Journal of Gambling Studies.* 23 (2007): 457–465.

Chapter 9: WEAPONS AND WEIRDNESS

Cai, Z., Richards, D. G., Lenhardt, M. L., and Madsen, A. G. "Response of human skull to bone-conducted sound in the audiometric-ultrasonic range." *International Tinnitus Journal.* 8 (2002): 3–8.

Cheney, M. *Tesla: Man out of time.* New York: Touchstone Books: New York, 2001.

Fletcher, N. H., Tarnopolsky, A. Z., and Lai, J. C. S. "Rotational aerophones." *Journal of the Acoustical Society of America.* 111 (2002): 1189–1196.

Foster, K. R., and Finch, E. D. "Microwave hearing: Evidence for thermoacoustic auditory stimulation by pulsed microwaves." *Science.* 185 (1974): 256–258.

Friedman, H. A. "U.S. PsyOp in Panama." http://www.psywarrior.com/PanamaHerb.html.

Friedman, H. A. "The Wandering Soul Psy-Op tape in Vietnam." http://www.psywarrior.com/wanderingsoul.html.

Gavreau, V. "Infrasound." *Science Journal.* 4 (1968): 33–37.

Gavreau, V., Condat, R., and Saul, H. "Infrasound: Generation, detection, physical properties and biological effects." *Acustica.* 17 (1966): 1–10.

Lebenthall, G. "What is infrasound?" *Progress in Biophysics and Molecular Biology.* 93 (2007): 130–137.

Lubman, D. "The ram's horn in Western history." *Journal of the Acoustical Society of America.* 114 (2003): 2325.

Liao Wanzhen. "Whistling arrows and arrow whistles." http://www.atarn.org/chinese/whistle/whistle.htm.

O'Brien, W. D. Jr. "Ultrasound—Biophysics mechanisms." *Progress in Biophysics and Molecular Biology.* 93 (2007): 212–255.

Tandy, V., and Lawrence, T. R. "The ghost in the machine." *Journal of the Society for Psychical Research.* 62 (1998): 1–7.

Chapter 10: FUTURE NOIZES

Bogatova, R. I., Bogomolov, V. V., and Kutina, I. V. "Trends in the acoustic environment at the International Space Station in the period of time of the ISS missions from one to fifteen." *Aerospace and Environmental Medicine.* 43 (2009): 26–30.

Doran, J., and Bizony, P. *Starman: The truth behind the legend of Yuri Gagarin.* New York: Walker & Co., 2011.

Elko, G. W. and Harney, K. P. "A history of consumer microphones: The electret condenser microphone meets microelectro-mechanical systems." *Acoustics Today.* 5 (2009): 4–13.

Hu, Z., and Corwin, J. T. "Inner ear hair cells produced in vitro by a mesenchymal-to-epithelial transition." *Proceedings of the National Academy of Sciences USA.* 104 (2007): 16675–16680.

Leighton, T. G., and Petculescu, A. "The sound of music and voices in space (Parts 1 & 2)." *Acoustics Today.* 5 (2009): 17–29.

Oshima, K., Shin, K., Diensthuber, M., Peng, A. W., Ricci, A. J., and Heller, S. "Mechanosensitive hair cell-like cells from embryonic and induced pluripotent stem cells." *Cell.* 141 (2010): 704–16.

Chapter 11: YOU ARE WHAT YOU HEAR

Buzsaki, G. *Rhythms of the brain.* New York: Oxford University Press, 2006.

INDEX

motion sickness, 208
motor-induced suppression, 193
movies, 106, 163–70
Mozart effect, 154–62
multicellular organisms, 18–19
multisensory integration, 101–2,
    183
music, 132–62, 277
    brain and, 160–61, 282–84
    Christmas, 149–51
    definition of, 139
    easy listening, 151–53
    emotions and, 145–46, 149–50,
        152–53
    in films, 164–70
    gamelan, 147–48
    habituation to, 150–53
    jingles, 179–85, 212
    mathematics of, 134–35
    mind control using, 203–7
    Mozart effect and, 154–62
    scientific research on, 135,
        138–49
    seismic recordings of, 135–38
    theme songs, 165–68
musical scale, 141
musical scores, 164–66
Muybridge, Eadweard, 164
Muzak, 151–53

nanoseconds, 85
negative emotions, 120–21,
    127–28
negative valence, 117–19
neural advertising, 212–15
neural imaging, 123–24
neural learning, 107, 116–17
neural prostheses, 248, 257
neurocinema, 164–65
neuroeconomics, 124, 211
neuroimaging, 280
neuronal filters, 275–76

neurons, 63, 89, 90, 96, 107–8,
    273–77, 283
neuron theory, 274–75
neuroscience, 101–2, 124
neurotransmitters, 60, 90, 107,
    192–93
noise
    background, 16, 40–41, 43,
        45, 217
    as constant, 1–4
    environmental, 150
    human-made, 22–23
    pink, 169
noise band, 40
noise-cancelling headphones,
    189
Noriega, Manuel, 230–31
nucleus accumbens, 123

octaves, 141, 149
operant conditioning, 184–85
Operation Just Cause, 230–31
opercularis pathway, 65, 68
optical illusions, 96
organisms
    multicellular, 18–19
    single-celled, 12–13
oscillogram, 37–39
ossicles, 49, 51
    incus, 49
    malleus, 49
    stapes, 49, 65
otolith organs, 14, 50–51, 65–66
    saccule, 50, 56, 65
    utricle, 65–66
outer ears, 74–77, 213–14
outer hair cells, 79
oxygen, 12

pain, 237
panning, 203
Paré, Denis, 121

voice, 41–42
volleys, 63
von Békésy, Georg, 78–79, 171
vox in vitro, 58

"The Wandering Soul" tape,
    228–30
water
    density of, 49–50
    hearing through, 49–52, 56, 58
    speed of sound in, 14, 50
Waterfall, Wallace, 171
Watson, Floyd, 171
wavelengths, 29–30
wave phenomena, modeling, 8–10
weapons, sonic, 218–45
Weberian ossicles, 51

Wernicke's region, 100, 129
Western clawed frog
    (*Xenopus tropicalis*), 55
whisper arches, 35
whistling arrows, 227
Williams, John, 166–67, 186
Wilson, Ben, 20
Worldwide Ear project, 258–60

Xenopus laevis, 54–60
Xenopus tropicalis, 55
The X-Files (TV show), 175–76

Yamaguchi, Ayako, 59–60

Zakon, Harold, 71
Zatorre, Robert, 161–62

A NOTE ON THE AUTHOR

Seth S. Horowitz, Ph.D., is a neuroscientist and an assistant research professor at Brown University. He is the cofounder of NeuroPop, the first sound design and consulting firm to use neurosensory and psychophysical algorithms in music, sound design, and sonic branding. He is married to sound and biomimetic artist China Blue and lives in Warwick, Rhode Island.